Renewable Energy and Biofuels

Renewable Energy and Biofuels

Edited by **Kurt Marcel**

New York

Published by Callisto Reference,
106 Park Avenue, Suite 200,
New York, NY 10016, USA
www.callistoreference.com

Renewable Energy and Biofuels
Edited by Kurt Marcel

International Standard Book Number: 978-1-63239-732-4 (Hardback)

Contents

Preface

Every book is initially just a concept; it takes months of research and hard work to give it the final shape in which the readers receive it. In its early stages, this book also went through rigorous reviewing. The notable contributions made by experts from across the globe were first molded into patterned chapters and then arranged in a sensibly sequential manner to bring out the best results.

This book discusses the fundamental as well as modern approaches and technologies related to the fields of renewable energy and biofuels. Renewable energy sources can be replenished in a given time frame and have the potential of replacing pollution causing fuels such as fossil fuels. Biofuels are derived from plants and other resources such as commercial, agricultural and domestic waste through biological processing using bacteria for anaerobic digestion. Biofuels are a commonly known form of renewable energy. The need to reduce the hazardous effect of non-renewable forms of energy has led to rapid progress and extensive research in this area. This extensive book collates researches from across the globe which explore the diverse aspects of renewable energy and biofuels. Different approaches, evaluations, methodologies and advanced studies have been included in this book. It will benefit researchers and students alike.

It has been my immense pleasure to be a part of this project and to contribute my years of learning in such a meaningful form. I would like to take this opportunity to thank all the people who have been associated with the completion of this book at any step.

Editor

Exploring the Relationship between Job Design and Knowledge Productivity: A Conceptual Framework in the Context of Malaysian Administrative and Diplomatic Officers

Mohamad Noorman Masrek[1], Nur Izzati Yusof[2], Siti Arpah Noordin[2] and Rusnah Johare[2]

[1]Accounting Research Institute/Faculty of Information Management, Universiti Teknologi MARA, Shah Alam, Malaysia

[2]Faculty of Information Management, Universiti Teknologi MARA, Shah Alam, Malaysia

Correspondence should be addressed to: Mohamad Noorman Masrek; mnoormanm@gmail.com

Abstract

To sustain its knowledge economic growth, Malaysian government had launched the K-Economy Master Plan in 2010; outlining the knowledge economy policy initiatives in supporting the knowledge based economy. The transition to knowledge based-economy, knowledge plays a governing role in leading the productivity and maintaining the economic performance growth. Equally important would be the factors that would contribute towards the knowledge creation. However, the literature indicates that very little attempts have been conducted to investigate factors that influence knowledge productivity. In Malaysia, the Administrative and Diplomatic Officers or PTD, are heavily involved in the development and implementation of the main policy makers of the federal government. However as to date, there are no mentions of any study exploring knowledge productivity of PTD. Against this background, this study intends to investigate the contributions of job characteristics towards knowledge productivity. In the process, it will propose a framework linking job characteristics and knowledge productivity. The proposed research method that will be employed will be a combination of qualitative and quantitative approaches. The qualitative approach will use interview as the main data collection technique, while the quantitative approach will adopt the questionnaire. The qualitative approach will help the researcher to develop the framework that depicts the factors that contribute towards knowledge productivity among PTD. Based on this framework a questionnaire will be developed so as to collect quantitative data involving PTDs. The contribution of the study can be viewed from two perspectives i.e. theoretical and practical. From the theoretical perspective, the study will develop an empirical based framework while from the practical perspective the study will help to devise strategies for improving and enhancing knowledge productivity. The significant of the study is that, it will further improve knowledge productivity among PTDs who are mainly involved in executing the country's development strategies including strengthening the administrative functions, social infrastructures and also the performance of the country's economic growth in line with the country's National Key Results Area (NKRA).

Keywords: Knowledge Productivity, Job Design, Conceptual Model, Malaysia.

Introduction

Knowledge productivity can be defined as the process in which an employee identifies, gathers, exchange and interprets relevant information and uses this information to develop new skills and applies the skills in innovating services and work processes. In developed countries such as United States of America, United Kingdom, France and Japan, knowledge has become of its main source of national income. Realizing the contribution and importance of knowledge productivity, the Malaysian government had launched the K-Economy Master Plan in 2002, outlining the knowledge economy policy initiatives in supporting the knowledge based economy. Since then, the government has placed strong attention on knowledge production activities as one of the critical means of generating national income. Various initiatives and policies have been developed and implemented to support and provide conducive climate for knowledge production activities.

An individual who is heavily involved in knowledge production is usually referred to as knowledge workers. A creative worker is an individual who is able to come up with creative ideas especially in the context of problem and strategy formulation. In an organizational setting, white collar jobs such as engineers and researchers are among those associated with knowledge worker. In the Malaysian public setting, the Administrative and Diplomatic Officer or PTD is also regarded as one of the most critical knowledge worker as their tasks and jobs heavily involved in the development of national level strategies and policies. Considering their nature of work, PTD plays very critical role in driving the country's economic activities. This is because the development and implementation of all government policies will affect all level society in the country ranging from an individual until giant business enterprise.

The literature suggests that various factors are linked or associated with knowledge production (e.g. Aminuddin, Tymms and Habsah, 2008; Sobia and Bakhtiar, 2011). These factors can be further categorized as organizational, individual and also external factors. Among these three groups of factors, the organizational and individual factors are considered the most influencing factors. The job design which is defined by Humphrey et al. (2007) as the elements of jobs and its environment, is considered as one of the strongest factors in shaping knowledge productivity. Job design by itself covers both organizational and individual factors. Given that very few studies have been done involving PTDs, very little is really known on the nature and scope of their jobs which have strong bearing on their knowledge production. Driven by this gap, this study attempts to investigate the nature of PTDs' job and how it affects their job productivity. As the main policy makers of the government, PTD will be the prime mover in promoting the widespread use of knowledge. Hence, the findings of the study is expected to further improve the knowledge productivity among PTDs who are mainly involved in executing the country's development strategies including strengthening the administrative functions, social infrastructures and also the performance of the country's economic growth in line with the government's National Key Results Area (NKRA).

Literature Review

Knowledge Productivity

Referring knowledge productivity as knowledge-based production process, Huang and Wu (2010) interpret knowledge productivity as "the capability with which individuals, teams, and units across an organization achieve knowledge-based improvements, exploitation, and innovations". On the other hand, Amiri, Ramezan and Omrani (2010) define knowledge productivity as "the learning ability in order to create knowledge-based results." Ramezan (2011) also defined knowledge productivity as "the way in which individuals, teams and units across an organization achieve knowledge-based improvements and innovations". In an attempt to measure knowledge production, researchers have developed various model and frameworks. In the process, the SECI model originally developed by Nonaka & Takeuchi (1995) has been referred and applied.

The SECI model proposed four ways that knowledge types can be combined and converted, showing how knowledge is shared and created in the organization. The four ways are (i) socialization (tacit to tacit) - knowledge is passed on through practice, guidance, imitation, and observation, (ii) Externalization (tacit to explicit) - tacit knowledge is codified into documents, manuals, etc. so that it can spread more easily through the organization, (iii) Combination (explicit to explicit) - codified knowledge sources (e.g. documents) are combined to create new knowledge, and (iv) internalization (explicit to tacit) - as explicit sources are used and learned, the knowledge is internalized, modifying the user's existing tacit knowledge.

Job Design

Various model and framework have been developed by various researchers to explain job design. The model developed by Morgeson & Humphrey (2006) is considered the most comprehensive model as it has been referred and applied by numerous researchers in studying job design of various professions (e.g. Humphrey et al., 2007 and Dere, 2011). The model describes work

design as consisting of motivational characteristics, knowledge characteristics, social characteristics and work context characteristics.

The work characteristic is further divided as consisting task autonomy, task variety, task significance, task identity and feedback from others. Task autonomy includes three interrelated aspects centered on freedom in (a) work scheduling, (b) decision making, and (c) work methods. Task variety refers to the degree to which a job requires employees to perform a wide range of tasks on the job. Task significance reflects the degree to which a job influences the lives or work of others, whether inside or outside the organization. Task identity reflects the degree to which a job involves a whole piece of work, the results of which can be easily identified. Feedback from job reflects the degree to which the job provides direct and clear information about the effectiveness of task performance.

According to Morgeson & Humphrey (2006), knowledge characteristics reflect the kinds of knowledge, skill, and ability demands that are placed on an individual as a function of what is done on the job. They further divided knowledge characteristics as consisting job complexity, information processing, problem solving, skill variety and specialization. Job complexity refers to the extent to which the tasks on a job are complex and difficult to perform. Information processing reflects the degree to which a job requires attending to and processing data or other information. Problem solving reflects the degree to which a job requires unique ideas or solutions and reflects the more active cognitive processing requirements of a job. Skill variety reflects the extent to which a job requires an individual to use a variety of different skills to complete the work. Specialization reflects the extent to which a job involves performing specialized tasks or possessing specialized knowledge and skill.

Within social characteristics, the job characteristics that involved include social support, interdependence, interaction outside organization and feedback from others (Morgeson & Humphrey, 2006). Social support reflects the degree to which a job provides opportunities for advice and assistance from others. Interdependence reflects the degree to which the job depends on others and others depend on it to complete the work. Interaction outside the organization reflects the extent to which the job requires employees to interact and communicate with individuals external to the organization. Feedback from others reflects the degree to which others in the organization provide information about performance.

Morgeson & Humphrey (2006) noted that work context characteristics include ergonomics, physical demands, work condition and equipment use. Ergonomics reflects the degree to which a job allows correct or appropriate posture and movement. Physical demands reflect the level of physical activity or effort required in the job. Work conditions reflect the environment within which a job is performed. Equipment use reflects the variety and complexity of the technology and equipment used in a job.

Job Design and Knowledge Productivity of PTD

The importance of job design in boosting individual's productivity and performance has been discussed extensively in few studies (Ali & Aroosiya, 2010; Fernando & Ranasinghe, 2010; Dere, 2011). From the organizational aspect, job design plays crucial part in supporting the employees' work performance in achieving organizational relevant outcomes as it may directly or indirectly influence the manner they perform their responsibilities and tasks (Ali & Aroosiya, 2010). Better known as Pegawai Tadbir dan Diplomatik or PTD, it is one of the positions serving under the public sector or government in Malaysia. The PTDs mainly involve in generalizing the country's

development strategies including strengthening the administrative functions, social infrastructures and also the performance of economic growths (Pegawai Tadbir dan Diplomatik, 1999).

PTD as the core knowledge worker in Malaysia have and will continue to act as the backbone of the knowledge economy initiatives. Their work diversity need to be in parallel towards the transition of knowledge based economy as outlined in the country's National Key Results Area (NKRA). As the main policy makers of the government, PTD will be the prime mover in promoting the widespread use of knowledge. Given the critical role played by PTD, study investigating their job design has been very limited. In addition, to date, there is still no study to measure their job productivity. As noted Morgeson & Humphrey (2006), study focusing on job design should be comprehensive and covers aspects such as motivational characteristics, knowledge characteristics, social characteristics and work context characteristics. In the same light, applicability of the SECI model which indicates the knowledge production process in the context of PTD is still unknown. In other words, a study is needed to unveil how the four processes namely socialization, externalization, combination and internalization take place in the context of PTD.

Conceptual Framework

Figure 1 illustrates the proposed framework for the study which is developed based on the work of Nonaka and Takeuchi (1995); Morgeson and Humphrey (2006); Humphrey, Nahrgang and Morgeson (2007) and Easa, (2012). The dependent variable for this framework is knowledge productivity which is derived from the work of Nonaka and Takeuchi (1995) and consists of four dimensions namely socialization, externalization, combination and internalization.

The independent variables of the framework are mainly adapted from Morgeson and Humphrey (2006) consists of the job design characteristics which are the motivational characteristics, social characteristics and work context characteristics. The dimensions of motivational characteristics are autonomy, task variety, task significance, task identity, feedback from job, job complexity, information processing, problem solving, skill variety and specialization. The dimensions of social characteristics are social support, interdependence, interaction outside organization and feedback from others. Whereas the dimensions of work context characteristics are ergonomics, physical demands, work conditions and equipment use.

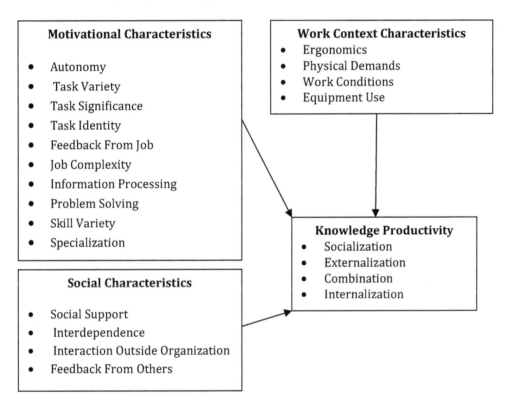

Fig. 1. Conceptual Framework

Research Methodology

Figure 2 is the illustration of the steps in research design in its entirety. There are two sources of identifying research topic, the literature and directly from the workplace or community settings (Gray, 2004). For the purpose of this study, both sources will be employed. Upon completion of formulating the research hypotheses, the researcher will conduct a preliminary study with a twofold objective. The first objective was to gain better understanding on the state of affairs of knowledge productivity among the PTDs. The

second objective is to clarify that the proposed research framework will reflect the real world phenomenon i.e. determining other potential factors that are applicable in the context with the present study. The findings obtained from the preliminary study are expected to help the researcher to refine the definitions of research problems, research objectives, theoretical framework and generation of research hypotheses.

The next stage of the study will involve in conducting a survey to collect the research data to be used for answering the research questions and testing the formulated hypotheses. The first part of the survey will involve in the development of the research instrument i.e. the survey questionnaire. In developing the questionnaires, validated measures that have been empirically used by previous researchers will be adapted and adopted. To ensure a quality and accurate research outcome resulting from the use of the research instrument, the questionnaire will undergo pre-testing and pilot study. The population of the study will be 5942 PTDs working in federal ministries located in Putrajaya. Following Gray (2004), the required sample size that should participate in this study would be 724. A stratified simple random sampling will be adopted. The population will be stratified according to ministries. Based on the PTD listing obtained from Public Service Department or Jabatan Perkhidmatan Awam (JPA), systematic random sampling will be used for each stratum. After obtaining a valid and satisfactory number of returned questionnaires, the researcher will then embark on data analysis and interpretation stage. From the results of data analysis and interpretation, deductions will be finally made.

Conclusion

PTD are directly involved in the Malaysian government policy making. The policies adopted by any country will surely have great impact to the society and economy of the country. Better and improved knowledge worker such as PTD will significantly influence their job quality and productivity which will also affect in the country's policy formulation. To date, studies investigating the factors influencing the knowledge productivity have received little attention in Malaysia. Thus, the conduct of this study will attempt to provide empirical evidences on the job design features that influence the knowledge productivity of PTD in Malaysia.

In order to develop the theoretical framework for the study, several theories and framework have been referred. The findings from the preliminary fieldwork will be applied to refine the theoretical framework to ascertain that it conforms to the actual phenomenon of job designs and knowledge productivity of the PTDs. Hence, the main contribution of the study is the establishment an empirical-based framework. The findings of the study can be used to either defy or strengthen the theories or frameworks that have been adapted. The instrument developed in the study can be used by researchers or practitioners to conduct further research in identifying the job design factors influencing the knowledge productivity in different settings or research design.

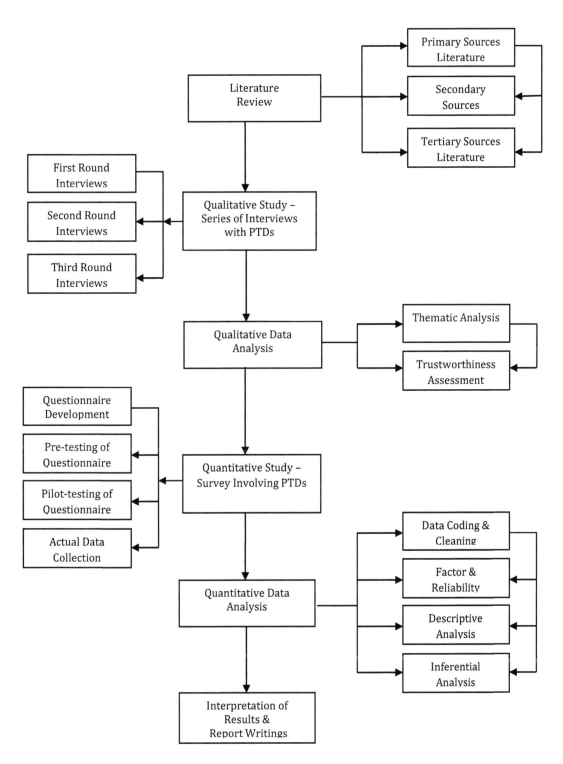

Fig. 2. Research Methodology

References

Ali, H. & Aroosiya, M. (2010). 'Impact of Job Design on Employees' Performance (With Sp. School Teachers in the Kalmunai Zone),' *Proceedings of the International Conference on Business & Information*, University of Kelaniya, Sri Lanka.

Amiri, A. N., Ramazan, M. & Omrani, A. (2010). 'Studying the Impacts of Organizational Organic Structure on Knowledge Productivity Effective Factors Case Study: Manufacturing Units in a Domestic Large Industrial Group,' *European Journal of Scientific Research*, 40(1), 91–101.

Anderson, P., Griego, O. V. & Stevens, R. H. (2010). 'Measuring High Level Motivation and Goal Attainment among Christian Undergraduate Students: An Empirical Assessment and Model,' *Business Renaissance Quarterly*, 5(1), 73-88

Dere, S. N. (2011). "A Diagnostic Exploration of Campus Recreation Using the Work Design Questionnaire," Unpublished Master's thesis, Texas State University. Retrieved from https://digital.library.txstate.edu/handle/10877/3341

Easa, N. F. (2012). "Knowledge Creation Process and Innovation in the Egyptian Banking Sector: Methodological Aspects," *Proceedings of the 20th EDAMBA Summer Academy*, Soreza, France.

Fernando, A. & Ranasinghe, G. (2010). 'The Impact of Job Design and Motivation on Employees Productivity as Applicable in the Context of Sri Lankan Software Engineers: A HR Perspective,' *Proceedings of the International Conference on Business & Information*, University of Kelaniya, Sri Lanka.

Gray, D. E. (2004). Doing Research in the Real World, London: *Sage Publications*.

Hassan, A., Tymms, P. & Ismail, H. (2008). "Academic Productivity as Perceived by Malaysian Academics," *Journal of Higher Education Policy and Management*, 30 (3), 283-296.

Huang, Y. C. & Wu, Y. C. J. (2010). "Intellectual Capital and Knowledge Productivity: The Taiwan Biotech Industry," *Management Decision*, 48 (4), 580 – 599.

Humphrey, S. E., Nahrgang, J. D. & Morgeson, F. P. (2007). "Integrating Motivational, Social, and Contextual Work Design Features: A Meta-Analytic Summary and Theoretical Extension of the Work Design Literature," *Journal of Applied Psychology*, 92, 1332-1356.

Morgeson, F. P. & Humphrey, S. E. (2006). "The Work Design Questionnaire (WDQ): Developing and Validating a Comprehensive Measure for Assessing Job Design and the Nature of Work," *Journal of Applied Psychology*, 91, 1321-1339.

Nonaka, I. & Takeuchi, H. (1995). The Knowledge-creating Company: How Japanese Companies Create the Dynamics of Innovation, *Oxford University Press*, New York.

Pegawai Tadbir dan Diplomatik. (1999). Retrieved from http://www.ptdportal.com/about_ptd/

Ramezan, M. (2011). "Examining the Impact of Knowledge Management Practices on Knowledge-Based Results," *Journal of Knowledge-based Innovation in China*, 3(2), 106 – 118.

Sobia, M. & Bakhtiar, A. (2011). 'An Empirical Investigation on Knowledge Workers Productivity in Telecom Sector of Pakistan,' *Information Management and Business Review*, 3 (1), 27-38.

The Criteria for Measuring Knowledge Management Initiatives: A Rare Glimpse into Malaysian Organizations

Reza Sigari Tabrizi[1], Yeap Peik Foong[1] and Nazli Ebrahimi[2]

[1]Multimedia University, Cyberjaya, Malaysia

[2]University of Malaya, KL, Malaysia

Abstract

Many challenges are facing measuring KM initiatives and one of the key challenges is to provide a comprehensive set of criteria to measure success of KM programs. The aim of this research is to address the problem of identifying the criteria for measuring KM outcomes among Malaysia companies and seeks to develop widely-accepted criteria based on the systematic review of the literature in order to measure success of knowledge management programs for Malaysian organizations. Hence, attempts were made to discover the most favored criteria among Malaysia organizations and to investigate the relationship between KM criteria and organization's mission, goals, and objectives. In addition, the relationship between KM criteria and success of KM programs were examined using regression analysis. The current population study was composed of 79 Malaysian organizations from different types of sectors. According to results achieved by statistical analyses, the most favored criteria among respondents who participated in this survey were enhanced collaboration, improved communication, improved learning/adaptation capability, sharing best practices, better decision-making, enhanced product or service quality, enhanced intellectual capital, and increased empowerment of employees. Finally, it is hoped that the current study provides a better picture for Malaysia organizations to identify and develop a comprehensive set of criteria to measure success of KM initiatives.

Keywords: Knowledge Management, Knowledge Management Outcomes, KM Criteria, Measuring KM Outcomes

Introduction

The current business environment is affected by a cutthroat competition, new launched products, and fast technology development (Davenport & Prusak, 1998). The backward-looking performance indicators are no longer sufficient since the knowledge era has begun and organizations need forward-looking indicators to move nimbly (Van Buren, 1999). According to Lubit (2001), today's core competencies and high performance have two primary bases, which are knowledge and intellectual capital. In fact, sustainability of competitive advantage that has derived from special knowledge inside companies is predominantly characterized by exhaustive competition among rivals and shortened product lifecycles (Lubit, 2001). Macintosh (1998) stated that exploiting knowledge assets of a company is a crucial issue to creating sustainable competitive advantage. Hence, Sustainability of companies' competitive advantage in chaos and uncertain business environment is highly related to implementing special knowledge to their core business processes and activities (Ndlela L. T. & du Toit, 2001).

Many organizations allocated such resources to implement knowledge management programs. However, latest research surveys have represented that despite companies have claimed to implement KM programs, not many of them are tagged as KM's successful implementer (Chong, Yew, & Lin, 2006). For the sake of implementing successful KM program, considering performance measurement is imperative and timely since not many organizations developed a well-organized performance measures to appraise their knowledge assets (Longbottom & Chourides, 2001). Hence, to organize a well-developed and formal performance measures is a crucial need for KM implementation within organizations (Chong, Yew, & Lin, 2006). In order to determine outcomes, structuring criteria for knowledge management efforts is an essential task of organization (Anantatmula & Kanungo, 2005). Needless to stress, the importance of determining criteria of measuring knowledge management efforts is significant.

Statement of the Problem

An important wide-accepted KM principle is a comprehensive set of criteria to measure outcomes of knowledge management efforts. It can be clearly seen that outcomes may not be identified without criteria; thus, structuring a set of criteria for knowledge management is imperative and timely (Chong, Yew, & Lin, 2006). Similar to a project or imitative that needs to meet a set of criteria to be selected; KM projects can also be evaluated through a set of criteria (Anantatmula & Kanungo, 2005). As such, companies have to establish metrics that are associated with KM criteria.

Knowledge Management Criteria

Perkmann (2002) investigated knowledge value from two different perspectives, which were the macro view and the micro view.

According to Perkmann (2002), the macro perspective measures intangible assets of a company by using means like Balance Scorecard, Score Board, Skandia navigators. The main advantage of macro perspectives is to evaluate knowledge management programs from non-financial approaches (Perkmann, 2002). In line with measuring knowledge value, Perkmann (2002) reported a measurement paradox of quantitative approaches. For example, it can be clearly seen that ROI as a financial ratio can only measure the financial gains of a specific project whereas there are many unintentional outcomes that may not be reflected by financial aspects. By contrast, Perkmann (2002) introduced a heuristic measure, which is named "Sveiby's Collaboration Climate Index" (CCI). The assumption behind the CCI is an excellent collaborative environment that facilitates knowledge sharing and hence increases organization's intellectual assets (Perkmann, 2002). Nonetheless, the CCI is a useful tool to find out the determinants, which are crucial for collaboration and knowledge sharing (Perkmann, 2002). In case of determining knowledge management outcomes, KPMG consulting (2000) has published a report on benefits of knowledge management program. KPMG (2000) conducted this research among 423 organizations in three different regions, which were United Kingdom, mainland Europe, United States.

Over 81 percent of the target organizations had knowledge management program, 38 percent had a KM program in place, 30 percent were preparing and 13 percent recognized the need to implement KM program (KPMG, 2000). Participants in KPMG (2000) research study indicated the percentage of the KM drivers inside organizations. According to KPMG (2000), 32 percent of board members, and 41 percent of senior management were belonged as knowledge management greatest drivers. This states that top management of companies supported knowledge management initiatives (KPMG, 2000). KPMG (2000) asked the respondents for their perspectives about the potential role of KM program that can contribute in gaining particular organizational goals. According to KPMG (2000), respondents believed that knowledge management program can play a role in achieving best results with respect to improving competitive advantage, marketing, improving customer focus, profit growth, product innovation, revenue growth, reducing costs, employee development, investment, and achieving mergers respectively.

BP AMOCO illustrated a set of parameters to assess knowledge management performance (Barrow, 2001). These parameters include efficient communication, employees' motivation, employees' morality, efficient knowledge sharing and transferring, efficient production management, effective project management, effective energy management, improving resource management, high product quality, high service quality, enhancing brand image, and improve company's efficiency (Barrow, 2001). Lynn, Reilly, and Akgün (2000) conducted a survey among such companies to find out the outcomes of knowledge management programs in new product teams. According to Lynn et al. (2000), the outcomes of knowledge management programs include cycle time reduction in launching new products, lower time-to-reach market, lower error and mistake in introducing new products, improving project documentation, more speed in retrieving information, efficient storage, access to best practices, and vision clearness.

Chong et al. (2006) exploited a list of KM outcomes that are grouped based on the previous works. According to Chong et al. (2006), outcomes can be incorporated into five different categories:

- Efficient Knowledge Processes

- Effective Personnel Development

- Customer Satisfaction

- Effective External Relationship

- Firm's Achievement

Knowledge process includes defining, creating, capturing, sharing, disseminating, and using knowledge assets (Van Buren, 1999). It needs to acquire personal knowledge to turn into organization's knowledge for sharing it through corporation (Chong et al., 2006). According to Chong et al. (2006), through systematic knowledge activity knowledge assets can be exploited effectively. One of the main objectives of knowledge management programs is to attract valuable experiences of knowledge workers (Chong & Choi, 2005). Today's high performance of organizations has two primary bases, which are knowledge and intellectual capital (Lubit, 2001). Ordonez de Pablos (2006) explained how intellectual capital relies on human, organizational, relational, and technological capitals. As Chong et al. (2006) stated, most valuable knowledge hold in employee's head, therefore, organizations are required to motivate their knowledge workers to share knowledge through commitment programs. Along with these programs, companies require to establish strong relationships with external environments involving suppliers and partners (Chong & Choi, 2005). Inside external zones, companies also need to acquire customer's experiences and knowledge (Van Buren, 1999).

Creating criteria for measuring knowledge management success is vital since criteria support to create a foundation for evaluating the value and assessing its outcomes (Anantatmula, 2005). In order to exploit criteria for evaluating knowledge management success, Anantatmula (2005) designed a questionnaire in which a list 26 KM outcomes was portrayed. The research targeted knowledge workers as respondents from various types of firms. The current research study adopted the questionnaire of Anantatmula.

Research Methodology

This section explains and discusses the systematic procedures that were performed in this survey.

Research Objectives

In this paper, an effort will be made to discover the criteria for measuring knowledge management success among Malaysian organizations. The focal objective of this study is to present criteria list that was adopted by Malaysian organizations to measure KM efforts. Specially, the following objectives were deployed to cover overall objectives of this paper.

• To ascertain the most favored criteria for measuring KM success

• To find out the dependency of the criteria on organization's mission, goals, and objectives

• To analyze the relationship between the criteria for measuring knowledge management results and the success of KM programs.

Research Questions

• What criteria are the most favored for measuring KM success?

• Are the criteria based on organization's mission, goals, and objectives?

• Is there any significant relationship between the criteria for measuring knowledge management results and the success of KM programs?

Hypotheses of the Study

The research hypotheses were depicted from research objectives as bellow:

• H_{10}: The criteria for measuring KM success are not dependent on mission, goals, and objectives.

• H_{11}: The criteria for measuring KM success are dependent on mission, goals, and objectives.

• H_{20}: There is no significant relationship between the criteria for measuring knowledge management results and the success of KM programs.

• H_{21}: There is a significant relationship between the criteria for measuring knowledge management results and the success of KM programs.

Data Analysis

In this research study, the SPSS software was used to analyze the questionnaire data. For this study, the proposed methods to find out hidden patterns were Descriptive Analysis, Multiple Regression Analysis, and Wilcoxon Signed Ranks Test.

Data Collection Method

For the purpose of this preliminary study, the following data collection method was used. This research study employed mixed-mode sampling approach in order of data collection. The first step of data collection was to choose a population to be sampled. The population framework was limited to web sites' forums, Yahoo discussion groups, Facebook discussion groups, email lists that have aggregated many different Malaysian executives, knowledge workers, knowledge management experts, and expats. Hence, generalizability across all Malaysian organizations is limited because of inherent constraints of the sample. Then, the online questionnaire was shared among all participants (Groups' members and email lists' contacts) and finally 79 of respondents answered the shared questionnaires. As expected, questionnaires were received with no missing variables under the population frame.

Participants

The participants of the survey's target population consist of KM professionals, Malaysian executives, and Expats executives who activated in Malaysia. These respondents were working in different types of organizations including Governmental, Non-governmental, For-profit, and Non-profit sectors. The questionnaire was developed on Google Document platform. The questionnaire then was shared with respondents using email lists and writing messages on their Social Networks' walls

Questionnaire

All surveys employ a questionnaire to collect relevant data. Questionnaires present a research instrument to collect information about employee's knowledge, motivations, mind-sets, and organizational behavior (Boynton & Greenhalgh, 2004). Questionnaire of Anantatmula provided a comprehensive list of KM Criteria, thus; the survey instrument in this research study was adopted from (Anantatmula, 2005). For this paper, all of the responses were collected using online questionnaire. The SPSS for windows version 16 was employed to generate summary outputs, graphs, and data analysis. The structure of the questionnaire was elaborated as bellow:

• The main objective of the questionnaire was to discover the criteria for measuring knowledge management success.

• The questionnaire consists of 19 questions including 16 close-ended questions as well as 3 open-ended questions.

• The questionnaire was divided into three sections, which were KM Criteria, Individual Background, and Organizational Background.

• In cover page, respondents were provided to get a brief explanation about the research topic.

• There was only one page that included all 26 criteria to arm the respondents' easiness to navigate between criteria and less time consuming to answer.

• In the last part of the questionnaire, respondents can give their email address to receive research findings.

• After submitting the online questionnaire, respondents can view latest summary of the survey.

Research Results

The statistical package employed for the survey data analysis was SPSS for Windows Version 16.0. Descriptive analysis was used to portray main attributes of the survey's data. Then, Wilcoxon Signed Ranks test was utilized to examine a hypothesis about the median of our target population. Finally, the KM criteria were regressed against success of KM programs using the Multiple Regression Analysis.

Demographic and Background Results

Types of Organizations

In the current survey, selected companies were activating in different types of organizations in Malaysia. As shown in Table 1, 53.16% of all organizations were operating as For-profit, 24.05% of which were operating as Non-Profit organizations. The remaining 22.78% were operating as Governmental organizations.

Operation Sectors of Organizations

The operation sectors of organizations were depicted in Table 2. Among the organizations investigated in this research study, 8.86% were operating in manufacturing sector. In addition, 30.38% of which were operating in Service industry, 21.52% are in Energy/Utilities, 1.27% are in Telecommunication, 15.19% are in Finance/ Banking/ Insurance, 5.06% are in Education, 8.86% are in R&D, and finally 8.86% are in trading sector.

Table 1: Types of Organizations

		Frequency	Percent	Valid Percent	Cumulative Percent
Valid	For-Profit	42	53.16	53.16	53.16
	Non-Profit	19	24.05	24.05	77.22
	Governmental	18	22.78	22.78	100
	Total	79	100	100	

Table 2: Operation Sectoprs of Organizations

		Frequency	Percent	Valid Percent	Cumulative Percent
Valid	Manufacturing	7	8.86	8.86	8.86
	Service	24	30.38	30.38	39.24
	Energy/Utilities	17	21.52	21.52	60.76
	Telecommunications	1	1.27	1.27	62.03
	Finance/Banking/Insurance	12	15.19	15.19	77.22
	Education	4	5.06	5.06	82.28
	R&D	7	8.86	8.86	91.14
	Trading	7	8.86	8.86	100
	Total	79	100	100	

Respondents' Role in Organizations

There were 79 participants to the survey, all of whom specified their role in their company. Table 3 represents respondents' role in organizations. As can be seen in Table 3, 13.92% of all respondents held position of CEO, 11.39% of whom held position of CIO/CKO, 15.19% were manager of HR, 26.58% were project manager, 21.52% project member and finally 11.39% of respondents held position of Professional Executive.

Table 3: Respondents' Role in Organizations

		Frequency	Percent	Valid Percent	Cumulative Percent
Valid	CEO	11	13.92	13.92	13.92
	CIO/CKO	9	11.39	11.39	25.32
	Manager of HR	12	15.19	15.19	40.51
	Project Manager	21	26.58	26.58	67.09
	Project member	17	21.52	21.52	88.61
	Professional Executive	9	11.39	11.39	100
	Total	79	100	100	

Table 4: Experience in Knowledge Management

		Frequency	Percent	Valid Percent	Cumulative Percent
Valid	1 to 2 years	19	24.05	24.05	24.05
	3 to 5 years	32	40.51	40.51	64.56
	6 to 10 years	24	30.38	30.38	94.94
	More than 10 years	4	5.06	5.06	100
	Total	79	100	100	

Experience in Knowledge Management

Table 4 represents the KM Experience gained by each participant during the years of working.

According to the above-tabulated results, 24.05% of all respondents had between 1 to 2 years experience, 40.51% of whom had between 3 to 5 years, 30.38% had between 6 to 10 years whereas only 5.06% of all respondents had more than 10 years experience in knowledge management.

Expertise in Knowledge Management

In this section, participants were asked to state their degree of expertise in knowledge management. The respondents' responses were illustrated in Table 5. According to Table 5, 20.25% of all respondents had Average level in KM, 24.05% of whom had above average whereas 55.7% of all respondents had excellent level of expertise in knowledge management.

Table 5: Expertise in Knowledge Management

		Frequency	Percent	Valid Percent	Cumulative Percent
Valid	Average	16	20.25	20.25	20.25
	Above average	19	24.05	24.05	44.3
	Excellent	44	55.7	55.7	100
	Total	79	100	100	

Analytical Results

Most Favored Criteria

Question 1 of the survey provided a list of 26 KM criteria. Participants were requested to clarify whether they have employed any of 26 criteria to measure knowledge management efforts in their companies or not. Respondents were also demanded to identify importance and effectiveness of each criterion based on the Likert scale. Both Importance and Effectiveness have equal Likert scale with 5 showing very high and 1 indicating very low. In order to calculate favored criteria, the mean scores of both Important and Effectiveness were computed for each criterion. Hence, the values nearer to 5 represent the most favored criteria. The list of favored scores for each criterion was represented in Table 6.

According to Table 6, a criterion with average of 3.85 or above can be considered as most favored criterion. As can be seen in Table 6, the most favored criteria include Enhanced collaboration (M=4.12, SD=1.02), Improved communication (M=4.07, SD=1.01), Improved learning/adaptation capability (M=3.94, SD=0.98), Sharing best practices (M=3.89, SD=0.95), Better decision making (M=3.89, SD=1.06), Enhanced product or service quality (M=3.89, SD=0.48), Enhanced intellectual capital (M=3.86, SD=1.01), and Increased empowerment of employees (M=3.85, SD=0.39).

KM Criteria and Mission, Objectives, and Goals

As noted in research methodology, H_1 examines the dependency of criteria for measuring knowledge management efforts on organization's mission, goals, and objectives. Hence, respondents were asked to assign a score to the dependency of criteria for measuring knowledge management success on organization's mission, goals, and objectives. The first step to examine the H_1 is to test the normality assumption. According to Royston (1992), the Shapiro-Wilk test is valid when sample size is greater than 3 and lesser than or equal to 2000. For this variable, the p-value for Shapiro-Wilk test of normality is 0.000, which is less than 0.05. Thus, the normality assumption was not met. Hence, the research hypothesis was tested using Wilcoxon Signed Ranks test. The Wilcoxon Signed Ranks test is applied in place of one-sample t-test when the normality assumption is not met (Chan, 2003). The results were represented in Table 7 and Table 8.

Table 6: The List of Criteria Based on Their Favored Rate

	N	Mean	Std. Deviation
Enhanced collaboration	79	4.1203	1.01973
Improved communication	79	4.0696	1.01190
Improved learning/adaptation capability	79	3.9430	.98380
Sharing best practices	79	3.8924	.95297
Better decision making	79	3.8924	1.05512
Enhanced product or service quality	79	3.8924	.48484
Enhanced intellectual capital	79	3.8608	1.00937
Increased empowerment of employees	79	3.8544	.39347
Improved productivity	79	3.7975	1.03316
Improved business processes	79	3.7848	1.08511
Improved employee skills	79	3.7152	.91876
New or better ways of working	79	3.7089	.85713
Return on investment of KM efforts	79	3.6456	.97452
Increased profits	79	3.6076	.90819
Better staff attraction/retention	79	3.5316	.93144
Better customer handling	79	3.4494	.91845
Improved new product development	79	3.4304	1.04922
Creation of more value to customers	79	3.2342	.69723
Faster response to key business issues	79	3.1899	1.13314
Increased innovation	79	3.1899	1.08988
Creation of new business opportunities	79	3.1329	.66847
Entry to different market type	79	3.0570	.63036
Increased market share	79	3.0316	.80599
Increased market size	79	2.9304	.94981
Reduced costs	79	2.8608	1.09760
Increased share price	79	2.6519	.58482
Valid N (listwise)	79		

Table 7: Table of Ranks in Wilcoxon Signed Ranks Test

		N	Mean Rank	Sum of Ranks
Hype_Mean - Criteria and Mission	Negative Ranks	64[a]	37.98	2430.50
	Positive Ranks	11[b]	38.14	419.50
	Ties	4[c]		
	Total	79		

a. Hype_Mean < Criteria and Mission

b. Hype_Mean > Criteria and Mission

c. Hype_Mean = Criteria and Mission

Table 8: Wilcoxon Signed Ranks Test

	Hype_Mean - Criteria and Mission
Z	-5.523[a]
Asymp. Sig. (2-tailed)	.000

a. Based on positive ranks.

b. Wilcoxon Signed Ranks Test

In this study, the test value was assumed equal to 3. According to Table 8, the p-value (Sig) equals to .000 which is less than 0.05; thus, the test would lead to reject H_{10} at level of $\alpha=0.05$. As shown in Table 7, most of the respondents would select 4 and 5 scores as their responses to this question. Therefore, the criteria for measuring knowledge management success are significantly based on organization's mission, goals, and objectives.

KM Criteria and Success of KM Programs Using Multiple Regression

The H_2 examines the relationship between the criteria for measuring knowledge management results and the success of KM programs. It is important to indicate that for Multiple Regression Analysis, the normality assumption should be tested. Therefore, the Shapiro-Wilk test was examined (3< n ≤2000). The Shapiro-Wilk statistics provided the p-value of 0.062, which was greater than 0.05. Thus, data can be assumed to be normally distributed. Hence, the Favored Criteria variables (See Section of Most Favored Criteria) were regressed against success of KM programs using stepwise Multiple Regression Analysis. The statement of "Do you think that knowledge management programs met the expected results?" was used to measure success of KM programs.

Favored Criteria and Success of KM Programs

The summaries of regression analysis were depicted in Table 9, 10, and 11. As shown in Table 9, SPSS generated four models. The model 4 was selected as final model to analyze the relationship between Success of KM programs as dependent variable and Favored Criteria as independent variables.

Table 9: - Model Summary - Criteria Favor on Meet Expected Results

Model	R	R Square	Adjusted R Square	Std. Error of the Estimate	Durbin-Watson
1	.840	.706	.702	.580	
2	.864	.747	.740	.542	
3	.875	.766	.756	.525	
4	.885	.783	.771	.509	1.984

From the Table 10, the F-value provided (F=66.590) which was significant at $\alpha=0.05$ (Sig=.000<0.05). This means that the regression model was fitted significantly and at least, one of the four independent criteria can be used to model success of KM programs. According to Table 9, the R-Square value produced ($R^2=78.3\%$). This indicated that 78.3 percent of variation in success of KM programs can be explained by all four independent variables. The Durbin-Watson of 1.984 falls between 1.5 and 2.5 (1.5<D-W<2.5) representing no autocorrelation among the error terms. Hence, it confirms that all error terms are independent.

The collinearity statistics indicate that tolerance statistics for Enhanced Intellectual Capital, Improved Productivity, Return on Investment of KM efforts, and Enhanced Product or Service Quality are all more than 0.1, and VIF (Variation Inflation Factors) are all lower than 10. Therefore, these show no multicollinearity problem.

Hence, H_2 was strongly supported and this represents that there is a significant relationship between the criteria for measuring KM results and the success of KM programs.

The results of Table 11 also confirmed that there were four criteria including Enhanced Intellectual Capital, Improved Productivity, Return on Investment of KM efforts, and Enhanced Product or Service Quality that were positively linked with success of KM programs. As can be seen in Table 11, the four criteria namely Enhanced Intellectual Capital (p<0.01), Improved Productivity (p<0.1), Return on Investment of KM efforts (p<0.05), and Enhanced Product or Service Quality (p<0.05) all directly contributed in the success of KM programs. Furthermore, the results also represented that the most important criteria that were involved in predicting success of KM programs was Enhanced Intellectual Capital and was statistically significant at $\alpha=0.01$ (p<0.01).

Table 10: ANOVA - Criteria Favor on Meet Expected Results

Model		Sum of Squares	df	Mean Square	F	Sig.
1	Regression	62.241	1	62.241	184.782	.000
	Residual	25.936	77	.337		
	Total	88.177	78			
2	Regression	65.866	2	32.933	112.183	.000
	Residual	22.311	76	.294		
	Total	88.177	78			
3	Regression	67.519	3	22.506	81.712	.000
	Residual	20.658	75	.275		
	Total	88.177	78			
4	Regression	69.006	4	17.252	66.590	.000
	Residual	19.171	74	.259		
	Total	88.177	78			

Table 11: Coefficients - Criteria Favor on Meet Expected Results [a]

Model		Unstandardized Coefficients		Standardized Coefficients			Collinearity Statistics	
		B	Std. Error	Beta	t	Sig.	Tolerance	VIF
1	(Constant)	.431	.260		1.661	.101		
	Enhanced intellectual capital	.885	.065	.840	13.593	.000	1.000	1.000
2	(Constant)	.276	.246		1.119	.267		
	Enhanced intellectual capital	.513	.122	.487	4.209	.000	.248	4.027
	Improved productivity	.419	.119	.407	3.514	.001	.248	4.027
3	(Constant)	.071	.253		.282	.779		
	Enhanced intellectual capital	.466	.120	.442	3.890	.000	.242	4.136
	Improved productivity	.313	.123	.305	2.545	.013	.218	4.585
	Return on investment of KM efforts	.216	.088	.198	2.450	.017	.477	2.096
4	(Constant)	1.363	.593		2.301	.024		
	Enhanced intellectual capital	.532	.119	.505	4.454	.000	.229	4.367
	Improved productivity	.212	.127	.206	1.674	.098	.194	5.160
	Return on investment of KM efforts	.224	.086	.205	2.617	.011	.476	2.099
	Enhanced product or service quality	.306	.128	.139	2.395	.019	.867	1.153

a. Dependent Variable: Meet Expected Results

Discussion of Findings

Based on the data collection from participants who were working for Malaysian organizations, effort was done to fulfill the objectives of this paper that is mainly, to determine the criteria for measuring knowledge management programs. As stated earlier, the accessibility of criteria as a platform to measure KM efforts would be delivering a great value to knowledge management programs inside organizations.

Most Favored Criteria

As shown in Table 6, the most favored criteria among respondents included: Enhanced collaboration (M=4.12, SD=1.02), Improved communication (M=4.07, SD=1.01), Improved learning/adaptation capability (M=3.94, SD=0.98), Sharing best practices (M=3.89, SD=0.95), Better decision making (M=3.89, SD=1.06), Enhanced product or service quality (M=3.89, SD=0.48), Enhanced intellectual capital (M=3.86, SD=1.01), and Increased empowerment of employees (M=3.85, SD=0.39). It can be clearly seen that establishing the measurements for these criteria needs critical thinking. Care must be taken that the intangible feature of above selected criteria makes it difficult to establish measurements for these criteria. For the sake of developing measures for some of the above favored criteria, Anantatmula (2005) proposed the following statements.

• Developing and promoting communication channels such as computer networks, organizational wiki pages, internal email system, and organizational social networks. This may help to develop a coherence transformation of employee's knowledge to organizational knowledge and vice versa.

• Establishing quantitative methods such as frequency of decision-making functions, and quantity of documented practices is a helpful procedure to measure communication aspect.

• Encouraging employees to contribute to organizational activities such as decision-making situations, and team working to solve management problems, is a valuable way to enhance collaboration inside organizations. It can be observed that the results and outputs of teams and committees are not relatively difficult to measure and evaluate.

Apart from above-mentioned solutions, companies can integrate some performance monitor tools with their network infrastructure to quantify number of shared organizations' practices, frequency of participation in workshops, seminars, problem solving committees, and quantity of achieved degrees and certifications. It can be also useful to provide feedback systems and suggestion box for measuring empowerment of employees (Anantatmula, 2005). Conducting organizational surveys to measure satisfaction and empowerment level of employees is another way to measure this criterion (Anantatmula, 2005). Finally, Total Quality Management as a strong instrument geared to ensure that company can measure the enhancing of product or service quality (Anantatmula, 2005).

KM Criteria and Organization's Mission, Goals and Objectives

According to literature review, criteria for measuring knowledge management efforts must associate and align with organizational mission, objectives, and goals. In this study, respondents were asked to give a score to their criteria depending on organizations' goals, mission, and objectives. According to the findings achieved from statistical analysis, the criteria for measuring knowledge management success were significantly based on organization's mission, goals, and objectives.

KM Criteria and Success of KM Programs

In order to analyze the relationship between KM Criteria and success of KM programs, the Favored Criteria variables were regressed against "Meet Expected Results" using

Stepwise Multiple Regression Analysis. According to the results achieved from Multiple Regression Analysis, a set of criteria that contributed in the success of KM programs were as bellow:

- Enhanced Intellectual Capital

- Improved Productivity

- Return on Investment of KM efforts

- Enhanced Product or Service Quality

All above-mentioned criteria have significant positive relationship with the success of knowledge management programs. Indeed, these criteria are aligned toward the success of KM efforts. The findings provided supporting evidence that success in KM efforts is highly dependent on developing measurement tools to evaluate these four criteria.

Limitations

Likewise each survey, this survey has its limitations some of which are; time restriction and budget constraint. These limitations as well as transportation problem compelled researchers to select a medium sample size. This is why researchers limited survey's population framework to email lists, Yahoo Discussion Groups, and Internet Forums etc. Hence, generalizability across all Malaysian organizations was limited because of inherent constraints of the sample. Furthermore, due to the above-mentioned limitations, this research study concentrated on only 26 KM criteria.

Recommendations for Future Researches

This study investigated the problem of determining the criteria to measure knowledge management initiatives among Malaysian firms. The results and findings can present viable and practical area of researches for future studies. The recommendations for future researches are stated as bellow:

- A study on the same topic with a larger pool of participants and a broad range of KM criteria.

- Break downing the most favored criteria to less abstract components in order to establish a clear measurement foundation for these criteria.

- Expanding the research to other countries in order of having multinational comparison.

- Developing research to special industry in order to get a better picture for investigation of that particular industry.

Conclusion

This paper attempted to determine criteria for measuring knowledge management success among Malaysian organizations. The major contribution of this study was to persuade managers to implement knowledge management programs toward organization's mission, goals, and objectives. Hence, defining well-organized and clear mission, goals, and objectives is an imperative task of top management. This may help organization to meet its expected results of KM programs. Analyzing the relationship between KM Criteria and the success of KM programs, led us to discover that by setting well-defined criteria and being aware of the importance of each criterion in measuring KM success, managers can adjust their programs on where they should spend their efforts and which area requires more concentration in order to get high achievement.

In conclusion, increasing the effectiveness of implementing KM programs and improving the quality of KM programs to satisfy the goals and the mission of the company will be the main value of the study, which can lead in gaining competitive advantage in current chaotic business environment.

Acknowledgement

We wish to express a sincere thank to Dr. VS Anantatmula who so graciously, agreed to use his questionnaire in this survey. We also would like to acknowledge the academic efforts of Dr. Chong Siong Choy in the knowledge management field.

References

Anantatmula, V. S. (2005). Outcomes of Knowledge Management Initiatives. *International Journal of Knowledge Management*, 50-67.

Anantatmula, V. & Kanungo, S. (2005). Establishing and Structuring Criteria for Measuring Knowledge Management Efforts. *38th Hawaii International Conference on System Sciences*, (pp. 1-11).

Barrow, D. C. (2001). Sharing Know-How At Bp Amoco. *Research-Technology Management*, 18-25.

Boynton, P. M. & Greenhalgh, T. (2004). Hands-on Guide to Questionnaire Research:Selecting, Designing, and Developing your Questionnaire. *BMJ*, 1312-1315.

Chan, Y. H. (2003). "Biostatistics 102:Quantitative Data – Parametric & Non-parametric Tests," Singapore Med J , 44 (8), 391-396.

Chong, S. & Choi, Y. (2005). Critical Factors of Knowledge Management Implementation Success. *Journal of Knowledge Management Practice*, 6 (6).

Choy, C. S., Yew, W. K. & Lin, B. (2006). Criteria for Measuring KM Performance Outcomes in Organisations. *Industrial Management & Data Systems*, 106 (7), 917-936.

Davenport, T. & Prusak, L. (1998). Working Knowledge: How Organisations Manage What They Know. Boston, Massachusetts: Harvard Business School Press.

KPMG. (2000). KM Articles: Knowledge Management Research Report. Retrieved February 15, 2010, from www.providersedge.com http://www.providersedge.com/docs/km_ar ticles/KPMG_KM_Research_Report_2000.pdf

Longbottom, D. & Chourides, P. (2001). Knowledge Management: a Survey of Leading UK Companies. Proceedings of the Second MAAQE International Conference, (pp. 113-26.). Versailles France.

Lubit, R. (2001). "Tacit Knowledge and Knowledge Management: The Keys to Sustainable Competitive Advantage,"*Organizational Dynamics*, 29 (4), 164–178.

Lynn, G. S., Reilly, R. R. & Akgün, A. E. (2000). Knowledge Management in New Product Teams:Practices and Outcomes. *IEEE Transactions on Engineering Management*, 47 (2), 221-231.

Macintosh, A. (1998). Position Paper on Knowledge Asset Management. Retrieved from Artificial Intelligence Applications Institute. http://www.aiai.ed.ac.uk/nalm/kam.html.

Ndlela, L. T. & du Toit, A. S. A. (2001). Establishing a Knowledge Management Programme for Competitive Advantage in an Enterprise. *International Journal of Information Management*, 21, 151-165.

Ordonez de Pablos, P. (2006). Transnational Corporations and Strategic Challenges An Analysis of Knowledge Flows and Competitive Advantage. *The Learning Organization*, 13 (6), 544-559.

Perkmann, M. (2002). Measuring Knowledge Value? Evaluating the Impact of Knowledge Projects. KIN brief.

Royston, P. (1992). Approximating the Shapiro-Wilk W-Test for Non-normality. *Statistics and Computing* 20:11-119.

Van Buren, M. E. (1999). A Yardstick for Knowledge Management. Training & Development, 71-78.

3

IT Enabled Engineering Asset Management: A Governance Perspective

Abrar Haider

School of Computer and Information Science, University of South Australia, Australia

Abstract

Engineering asset lifecycle management requires a variety of information as well as operational technologies to keep their asset base in running condition In theory these technologies are used in collection, storage, and analysis of information spanning asset lifecycle processes; providing decision support capabilities through analytic conclusions arrived at from analysis of data; and in providing an integrated view of asset management through processing and communication of information that also allows for the basis of asset management functional integration. In doing so, these technologies not only provide for the control of asset lifecycle tasks, but also contribute to the overall advise on effective asset management though the critical role that they have in decision making. However, even though operational technologies depend a lot on information technologies for their smooth functioning, yet due to their specialized nature these operational technologies are not considered as part of the overall organizational information technology infrastructure. Consequently, when it comes to governance of information technologies, operational technologies are not accounted for. This paper provides a framework for governance of information technologies utilized for asset lifecycle management. It concludes that information technologies should not be taken as technical constructs, these are at the core of strategic alignment, value delivery, resource management, and risk management. Governance of information technology, therefore, calls for understanding and accounting for the whole information technology base and enabling infrastructure of the organization.

Keywords: Asset management, IT governance, Asset lifecycle

Introduction

Information Technologies (IT) for asset management are required to translate strategic objectives into action; align organizational infrastructure and resources with IT; provide integration of lifecycle processes; and inform asset and business strategy through value added decision support. However, the fundamental element in achieving these objectives is the quality of alignment of technological capabilities of IT with the organizational infrastructure, as well as their fit with the operational technologies (OT) used in lifecycle management of assets. IT and OT are becoming inextricably intertwined, where OT facilitate running of the assets and is used to ensure system integrity and to meet the technical constraints of the system. OT includes control as well as management or supervisory systems, such as SCADA, EMS, or AGC. These systems not only provide the control of asset lifecycle tasks, but also contribute to the overall advice on effective asset management though the critical role that they have in decision making. However, even though OT owes a lot to IT for their smooth functioning, yet due to their specialized nature these technologies are not considered as IT infrastructure. This paper, therefore, attempts to uncover the relationship between industry specific OT used for asset management and organizational use of mainstream IT applications for asset lifecycle management. It starts with an analysis of the IT utilized for asset management, which is flowed by a discussion on their relationship with OT in asset lifecycle management. The paper, thus, presents a framework for IT-OT nexus.

Asset Management

The scope of asset management activities extends from establishment of an asset management policy and identification of service level targets according to the expectation of stakeholder and regulatory/legal requirements, to the daily operation of assets aimed at meeting the defined levels of service. Asset managing organizations, therefore, are required to cope with the wide range of changes in the business environment; continuously reconfigure manufacturing resources so as to perform at accepted levels of service; and be able to adjust themselves to change with modest consequences on time, effort, cost, and performance.

Asset management can be classified into three levels, i.e. strategic, tactical, and operational. Strategic level is concerned with understanding the needs of stakeholders and market trends, and linking of the requirements thus generated to the optimum tactical and operational activities. Operational and tactical levels are underpinned by planning, decision support, monitoring, and review of each lifecycle stage to ensure availability, quality, and longevity of asset's service provision. The identification, assessment, and control of risk is a key focus at all levels of planning, with the results from this process providing inputs into the asset management strategy, policies, objectives, processes, plans, controls, and resource management.

IT and Asset Management

In theory IT in asset management have three major roles; firstly, IT are utilized in collection, storage, and analysis of information spanning asset lifecycle processes; secondly, IT provide decision support capabilities through the analytic conclusions arrived at from analysis of data; and thirdly, IT provide an integrated view of asset management through processing and communication of information and thereby allow for the basis of asset management functional integration. According to Haider (2007), minimum requirements for asset management at the operational and tactical levels are to provide functionality that facilitates, knowing what and where are the assets that the organization owns and what is their condition; establishing suitable maintenance, operational and renewal regimes to suit the assets and the level of service required of them by present and future customers; implementing job/resources management, and improving risk management techniques; and identifying the true cost of operations and maintenance; and optimizing operational procedures.

In engineering enterprises asset management strategy is often built around two principles, i.e., competitive concerns and decision concerns (Rudberg, 2002). Competitive concerns set manufacturing/production goals, whereas decision concerns deal with the way these goals are to be met. IT provide for the these concerns through support for value added asset management, in terms of the choices such as, selection of assets, their demand management, support infrastructure to ensure smooth asset service provision, and process efficiency. Furthermore, these choices also are concerned with in-house or outsourcing preferences, so as to draw upon expertise of third parties. IT not only aids in decision support for outsourcing of lifecycle processes to third parties, but also provide for the integration of extra-organizational processes with the intra-organizational processes. Nevertheless, the primary expectation from IT at the strategic level is that of an integrated view of asset lifecycle, such that informed choices could be made in terms of economic tradeoffs and/or alternatives for asset lifecycle in line with asset management goals, objectives, and long term profitability outlook of the organization. However, according to IIMM (2006), the minimum requirements for asset management at the strategic level are to aid senior management in,

a. predicting the future capital investments required to minimize failures by determining replacement costs;

b. assessing the financial viability of the organization to meet costs through estimated revenue;

c. predicting the future capital investments required to prevent asset failure;

d. predicting the decay, model of failure or reduction in the level of service of assets or their components, and the necessary rehabilitation/ replacement programmers to maintain an acceptable level of service.

e. assessing the ability of the organization to meet costs (renewal, maintenance, operations, administration and profits) through predicted revenue;

f. modelling what if scenarios such as, technology change/obsolesce; changing failure rates and risks they pose to the organization, and alterations to renewal programs and the likely effect on levels of service,

g. alteration to maintenance programs and the likely effect on renewal costs; and

h. impacts of environmental (both physical and business) changes.

IT for asset management seeks to enhance the outputs of asset management processes through a bottom up approach. This approach gathers and processes operational data for individual assets at the base level, and on a higher level provides a consolidated view of entire asset base (figure 1

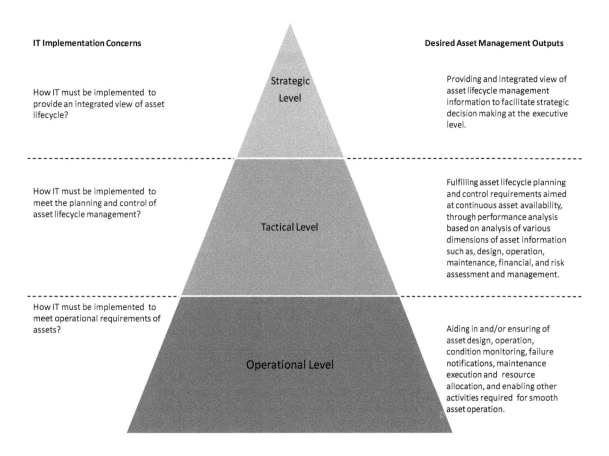

IT Implementation Concerns

How IT must be implemented to provide an integrated view of asset lifecycle?

How IT must be implemented to meet the planning and control of asset lifecycle management?

How IT must be implemented to meet operational requirements of assets?

Strategic Level

Tactical Level

Operational Level

Desired Asset Management Outputs

Providing and integrated view of asset lifecycle management information to facilitate strategic decision making at the executive level.

Fulfilling asset lifecycle planning and control requirements aimed at continuous asset availability, through performance analysis based on analysis of various dimensions of asset information such as, design, operation, maintenance, financial, and risk assessment and management.

Aiding in and/or ensuring of asset design, operation, condition monitoring, failure notifications, maintenance execution and resource allocation, and enabling other activities required for smooth asset operation.

Figure 1: Scope of IT for Asset Management (Haider 2009)

At the operational and tactical levels, IT systems are required to provide necessary support for planning and execution of core asset lifecycle processes. For example, at the design stage, designers need to capture and process information such as, asset configuration; asset and/or site layout design and schematic diagrams/drawings; asset bill of materials; analysis of maintainability and reliability design requirements; and failure modes, effects and criticality identification for each asset. Planning choices at this stage drives future asset behavior, therefore the minimum requirement laid on IT at this stage are to provide right and timely information, such that informed choices could be made to ensure availability, reliability and quality of asset operation. An important aspect of asset design stage is the supportability design that

governs most of the later asset lifecycle stages. The crucial factor in carrying out these analyses is the availability and integration of information, such that analysis of supportability of all facets of asset design and development, operation, maintenance, and retirement are fully recognized and defined. Nevertheless, effective asset management requires the lifecycle decision makers to identify the financial and non financial risks posed to asset operation, their impact, and ways to mitigate those risks.

OT and Asset Management

OT set of technologies are primarily used for process control; however, they also include technologies such as sensors, gauges, and meters, which are used in many control systems and automated data acquisition systems that perform a variety of tasks within the asset lifecycle. Technically, OT is a form of IT as it necessarily deals with information and is controlled by (in most cases) a software. For example, the Supervisory Control and Data Acquisition (SCADA) systems used for real time monitoring and control of processes consist of software and hardware and produces intelligible information that is used for a variety of follow up actions and decision support.

From the discussion on IT and OT for asset management, it is clear that these technologies not only have to provide for standardized quality information but also have to provide for the control of asset lifecycle processes. For example, design of an asset has a direct impact on its asset operation. Operation, itself, is concerned with minimizing the disturbances relating to production or service provision of an asset. At this level, it is important that IT systems are capable of providing feedback to maintenance and design functions regarding factors such as asset performance; detection of manufacturing or production process defects; design defects; asset condition; asset failure notifications. There are numerous IT systems employed at this stage that capture data from sensors and other field devices to diagnostic/prognostic systems; such as SCADA systems, Computerized Maintenance Management Systems (CMMS), and Enterprise Asset Management systems. These systems further provide inputs to maintenance planning and execution. However, effective maintenance not only requires effective planning but also requires availability of spares, maintenance expertise, work order generation, and other financial and non financial supports. This requires integration of technical, administrative, and operational information of asset lifecycle, such that timely, informed, and cost effective choices could be made about maintenance of an asset. For example, a typical water pump station in Australia is located away from

major infrastructure and has considerable length of pipe line assets that brings water from the source to the destination. The demand for water supply is continuous for twenty four hours a day, seven days a week. Although, the station may have an early warning system installed, maintenance labour at the water stations and along the pipeline is limited and spares inventory is generally not held at each station. Therefore, it is important to continuously monitor asset operation (which in this case constitutes equipment on the water station as well as the pipeline) in order to sense asset failures as soon as possible and preferably in their development stage. However, early fault detection is not of much use if it is not backed up with the ready availability of spares and maintenance expertise. The expectations placed on water station by its stakeholders are not just of continuous availability of operational assets, but also of the efficiency and reliability of support processes. IT and OT systems, therefore, need to enable maintenance workflow execution as well as decision support by enabling information manipulation on factors such as, asset failure and wear pattern; maintenance work plan generation; maintenance scheduling and follow up actions; asset shutdown scheduling; maintenance simulation; spares acquisition; testing after servicing/repair treatment; identification of asset design weaknesses; and asset operation cost benefit analysis. An important measure of effectiveness of IT and OT, therefore, is to treat operational technologies as information technologies are governing them with the same guidelines as the overall IT infrastructure is managed. An integrated governance framework of IT and OT will allow for setting up a regime that will provide standardisation of quality and interoperable information through development and procurement of appropriate hardware and software applications; establishing appropriate skill set of employees to process information; and the strategic fit between the asset lifecycle management processes and technology.

Governance of IT Based Asset Management

IT resources represent the combination of IT infrastructure, human IT resources, and the soft assets involved in the use of IT (Gunasekaran et al., 2006), such as the shared performance and prospect development potential of an organisation (Lin, 2007). Implementation of these technologies should, therefore, properly match the process requirements. Implementation considerations need to account for internal development of the organisation as well as addressing the external forces impacting the organisation. Organisations improve externally and internally by making decisions which may affect the learning, acquiring and operation of IT resources (Stoel and Muhanna, 2009). The closeness between the CEO and CIO can improve the organisation by bringing new technology and supporting organisational changes, which are vital for achieving internal efficiencies as well for competitiveness of the organisation (Ranganathan and Kannabiran, 2004; Booth and Philip, 2005). It is therefore, important to have appropriate governance structures in place that treat IT infrastructure and related resources as strategic assets and guide the organization on achieving internal as well as external efficiencies through the use of IT.

There are many definitions of IT governance in the extant literature. Some researchers argue that IT governance is the organisational capability operated by the board, executive management and IT management to organize the creation and implementation of IT strategy to certify the combination of business and IT (Grembergen, 2004). However, IT Governance Institute (2005) describes it as the accountability of the leadership and posits that it is a fundamental component of Corporate Governance which involves the management and organisational structures and processes to certify that the organisation's IT maintain and broaden the organisation's strategies and objectives.

Luftman (1996) in Grembergen (2004) contends that IT governance is the extent to the rights for IT decision-making which is determined and shared between management and the processes of leadership in both IT and business enterprises that consists of IT priorities and IT resources distribution. These definitions show that the issues of IT governance has been approached and investigated by researchers from a variety of angles. However, this research accepts that IT governance is the decision rights and accountability framework for encouraging desirable outcomes and behaviours in the use of IT (Weil and Ross, 2004). In crux, IT governance addresses the organizational resources which control IT infrastructure, execute IT strategy, and ensure business IT assets fit with the business strategy (Brown, 2006). It embodies strategic information system planning and management, ensuring system reliability through internal controls, and managing-system related business risks (O'Donnell, 2004). IT governance involves the relationship between IT and business management by combining business systems thinking, which concerns business knowledge and understanding of IT to support the relationships and skills of employees in both business and IT areas (Liu, Lu and Hu, 2008). The five core areas of IT governance include value delivery, risk management, performance management, resources management, and strategic alignment. IT governance, thus, allows an organization to achieve three important objectives, which are decision-making, functional superiority, and risk management optimization. There are a variety of potential frameworks which may be suitable to apply or implement in organisations and different industries. IT governance is strongly influenced by factors such as company size, expansion forecasts, business processes, IT operations, industry, financial health of the organisation, and IT support infrastructure (Dehning, Richardson and Stratopoulos, 2005). However, the success of a governance framework depends upon aligning business goals and IT operational processes to deliver value, IT strategy, and build internal efficiencies; through effective audit, control and management of IT and related resources

in diverse business aspects such as operation, compliance, finance and IT risk (Tuttle and Vandervelde, 2007).

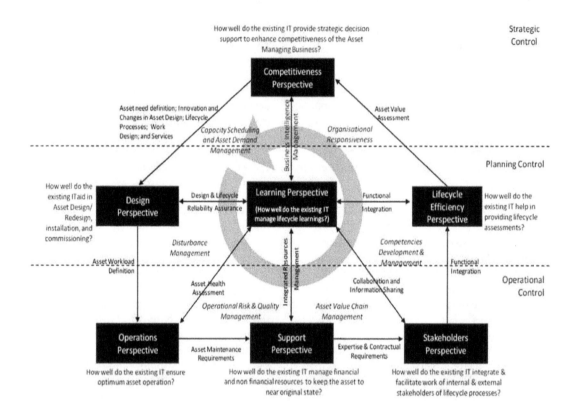

Figure 2: Technology Governance Framework for Asset Management

Figure 2 illustrates an IT based engineering asset governance framework. It is a learning centric framework and accounts for the core asset management processes as well as the allied areas where IT make contributions. It therefore accounts for the soft as well as the hard benefits gained from IT utilisation in an asset lifecycle.

This framework divides the asset lifecycle into 7 perspectives, where each perspective consists of processes that contribute to asset lifecycle management. The framework begins with assessing the usefulness and maturity of IT in mapping the organisation's competitive priorities into asset design and reliability

support infrastructure. The framework thus assesses the contribution and maturity of IT through four further perspectives before informing the competitive priorities of the asset managing organisation. In so doing, the framework translates asset management strategy into action through the use of IT. At the same time, this framework could be used as an evaluation framework to examine the role of IT as strategic translators as well as strategic enablers of asset lifecycle management and enables generative learning. It means that instead of just providing a gap analysis of the desired versus actual state of IT maturity and contribution, it also assesses the information requirements at each perspective and thus enables continuous improvement through action oriented evaluation learning.

Capacity and Demand Management

In a usual asset lifecycle, asset demand and capacity specify the nature of assets, as well as the types of supportability infrastructure required to ensure asset reliability through its lifecycle. The success of IT at this stage depends upon the availability, speed, depth, and quality of information regarding competitive environment of the organisation. This information allows asset managers to measure the demands of asset customers, which specifies the types of assets or the improvements required in existing asset configuration to address the customers' demands. At this stage, asset managers require the IT to provide them with decision support capabilities by accounting for economic and environmental constraints, optimised levels of asset utilisation, and costs of asset reliability to ensure sustainable service delivery. The nature of this information is multifaceted and therefore, requires scanning of the external business environment as well as taking into consideration the learnings gained over the years from managing assets employed by the organisation.

The value profile that asset managers attach to IT at this point, is of business intelligence management, so as to aid the design of the asset as well as the support infrastructure. Within design perspective itself, there is a variety of information demands that the IT are required to fulfil. In a nutshell, the value profile of IT demanded by the asset designers specifies how the IT aid in asset design/re-design, installation, and commissioning. Nevertheless, each of these processes further consist of a series of activities that require an assortment of information to enable evaluations and alternative solutions, such that the organisation is able to choose the best possible solution to asset design/redesign. These alternatives are arrived at after having considered a series of

analysis that encompass the capability potential and associated costs for ensuring reliability of the asset operation. The success factor of IT in ensuring asset supportability and design reliability is the depth and coverage of supportability analysis, which

provide a roadmap for the later stages of the asset lifecycle. These analyses not only specify the costs associated with supporting the asset lifecycle, but also identify other critical aspects such as the throughput of the asset, spares requirements, and training requirements. Therefore, at this stage it is important to assess how IT meet the demands of asset design and design for supportability of asset reliability, as well as their integration with other IT in the organisation and the capacity of IT to preserve learnings and make them available throughout the organisation.

Disturbance Management

Asset workload is defined according to its 'as designed' capabilities and capacity. However, during its operational life, every asset generates some maintenance demands. During the asset operation stage, the critical feature of IT is to aid asset managers in managing disturbances. This requires availability of design as well as supportability information, as well as current information on the condition of an asset. Different organisations deploy different condition or health monitoring systems, such as sensors, manual inspections, and paper based systems. Nevertheless, IT at this stage need to be able to provide consolidated health advisories by capturing and integrating this information, analysing asset workload information, health information, and design information to enable speedy malfunction alarms and communication of failure condition information to maintenance function. Many of the design errors surface during asset operation, therefore, it is also important to assess if the existing IT systems report back these errors to the asset design function so as to ensure asset design reliability. At the same time, it is important to assess the contribution of IT in enabling asset lifecycle processes under this perspective, along with the level of IT integration, and the contribution that they make in preserving lifecycle learnings.

Operational Risk Management

The notion of risk signifies the 'vulnerabilities' that asset operation is exposed to, due to operating in a particular physical setting or specific work conditions. Nevertheless, the success of risk management is dependent upon factors such as availability of expertise to carry out maintenance treatments, availability of spares, maintenance expertise, maintenance project management as well as complete information on the health status and pervious maintenance history of the asset. The role of IT therefore needs to be assessed for their ability to provide control of decentralised tasks and to ensure the availability of resources to keep the assets in near original state. However, as with the previous sections, the significant factor is to preserve the learnings from maintenance execution and making the same available to other functions of asset lifecycle so as to enable holistic decision support regarding asset maintenance, renewal, and retirement.

Asset Operation Quality Management

The aim of asset managing processes is to keep the asset to or near its original or as designed state throughout its operational life. Therefore, once a disturbance has been identified, it becomes crucial to curtail its impact to minimum and to take appropriate follow up actions. These follow up actions not only involve the direct actions taken on the asset such as maintenance execution, but also involve sourcing of maintenance, rehabilitation, and renewal materials and expertise as well as the contractual agreements. At the same time with the growing attention being given to the environment, it is equally important to ensure that the asset operation conforms to the governmental and industrial regulations,

and to control the impact of disturbance on the environment. IT at this stage have a versatile role, and aid in maintenance and rehabilitation execution, enabling collaboration and communication, managing resources, as well as facilitating business relationships with external stakeholders and business partners. It is therefore important to measure these value provisions of IT at this stage.

Competencies Development and Management

During the course of performing asset lifecycle management activities, engineering organisations generate enormous amount of explicit as well as tacit knowledge. The knowledge thus generated, provides an organisation with competencies in managing its assets. IT not only have the ability to capture and process this knowledge, but can also facilitate knowledge sharing among organisational stakeholders. However, in order for this to happen, it is important to find the fit between the social and technical systems in the organisation, since competencies development depends upon the functional/technical knowledge, as well as cultural, social, and personal values.

Organisational Responsiveness

Functional integration and a consolidated view of the asset lifecycle facilitate the asset managing organisation in responding to the internal as well as external changes. IT play an important role in materialising such responsiveness, due mainly to their ability to provide asset lifecycle profiling from financial and non financial perspectives. These value assessments aid the organisation in making decisions, such as asset redesign, retirement, renewal, as well as cost benefits of service provision and asset operation, and assessments of market demands. Nevertheless, the fundamental requirements in producing these value assessments are the availability integrated and quality information that allow for an integrated view of asset lifecycle though maintaining the asset lifecycle learnings. This framework enables action oriented learning, as it highlights the gaps between the existing and desired levels of performance, thereby necessitating the need for corrective action through (re)investment in right technology and skills, and acceptance of the change in the organisation. The evaluation thus provides triggers for continuous improvement regarding IT employed for asset design, operation, maintenance, risk management, quality management, and competencies development for asset lifecycle management.

Conclusion

IT utilised in asset management not only have to provide for the decentralized control of asset management tasks but also have to act as instruments for decision support. However, information requirements for control and decision support in asset lifecycle management are prone to change, due mainly to the changes in the business, operational, and environmental environment. The ability of an organisation to understand these changes not only contributes to its responsiveness, but also improves its capacity to enhance reliability of asset operations, to deliver optimised level of asset lifecycle management. However, this ability is directly influenced by the way an organisation governs its IT infrastructure, which consequently acquires, processes, and presents information to enable asset managing organisations to understand these changes. This paper has presented a governance framework for IT utilised in engineering asset lifecycle management. This framework translates strategic objectives into action; aligns organisational infrastructure and resources with information technology and related resources; providing integration of lifecycle processes; and ensures informing asset and business strategy through value added decision support.

References

Alexander, K. (2003). "A Strategy for Facilities Management," *Facilities*, 21 (11/12), 269 – 274.

Balch, W. F. (1994). "An Integrated Approach to Property and Facilities Management," *Facilities*, 12 (1), 17-22.

Booth, M. E. & Philip, G. (2005). "Information Systems Management in Practice: An Empirical Study Of UK Companies," *International Journal of Information Management*, 25 (4), 287 – 302.

Boyle, T. A. (2006). "Towards Best Management Practices for Implementing Manufacturing Flexibility," *Journal of Manufacturing Technology Management*, 17 (1), 6-21.

Brown, W. C. (2006). "IT Governance, Architectural Competency, And The Vasa," *Information Management & Computer Security*, 14 (2), 140 - 154.

Dehning, B., Richardson, V. J. & Stratopoulos, T. (2005). "Information Technology Investments and Firm Value," *Information & Management*, 42 (7), 989 – 1008.

El Hayek, M., Voorthuysen, E. V. & Kelly, D. W. (2005). "Optimizing Life Cycle Cost of Complex Machinery With Rotable Modules Using Simulation," *Journal of Quality in Maintenance Engineering*, 11 (4), 333-347.

Gottschalk, P. (2006). "Information Systems in Value Configurations," *Industrial Management and Data Systems*, 106 (7), 1060-1070.

Grembergen, W. V. (2004). Strategies for Information Technology Governance, *IDEA Group Publishing*, United States of America and United Kingdom.

Gunasekaran, A., Ngai, E. W. T. & McGaughey, R. E. (2006). "Information Technology and Systems Justification: A Review for Research and Applications," *European Journal of Operational Research* , 173 (3), 957 - 983.

Haider, A. (2007). "Information Systems Based Engineering Asset Management Evaluation: Operational Interpretations," *PhD Thesis*, University of South Australia, Adelaide, Australia.

Haider, A. (2009). 'Value Maximisation from Information Technology in Asset Management – A Cultural Study,' Proceedings of the International Conference of Maintenance Societies (ICOMS), 2-4 June 2009, Sydney, Australia.

IIMM (2006). 'International Infrastructure Management Manual,' Association of Local Government Engineering NZ Inc, National Asset Management Steering Group, New Zealand, Thames, ITBN 0-473-10685-X.

Inman, R. A. (2002). "Implications of Environmental Management for Operations Management," *Production Planning and Control*, 13 (1), 47-55.

IT Governance Institute, (2005). "IT Governance Domain Practices and Competencies: Optimising Value Creation-From IT investments," IT Governance Institute. [Online], [Retrieved March 15, 2010], http://www.isaca.org/ContentManagement/ContentDisplay.cfm?ContentID=33923.

Lin, B. (2007). "Information Technology Capability and Value Creation: Evidence from The US Banking Industry,"*Technology in Society*, 29 (1), 93 – 106.

Liu, Y., Lu, H. & Hu, J. (2008). "IT Capability as Moderator Between IT Investment and Firm Performance," *Tsinghua Science & Technology* , 13 (3), 329-336.

Murthy, D. N. P., Atrens, A, & Eccleston, J. A. (2002). "Strategic Maintenance Management," *Journal of Quality in Maintenance Engineering* , 8 (4), 287-305.

Narain, R., Yadav, R. C., Sarkis, J. & Cordeiro, J. J. (2000). "The Strategic Implications of Flexibility in Manufacturing Systems," *International Journal of Agile Management Systems*, 2 (3), 202-13.

O'Donnell, E. (2004). "Discussion of Director Responsibility for IT Governance: A Perspective on Strategy," I*nternational Journal of Accounting Information Systems*, 5 (2), 101-104.

Ranganathan, C. & Kannabiran, G. (2004). "Effective Management of Information Systems Function: An Exploratory Study of Indian Organizations," *International Journal of Information Management*, 24 (3), 247 – 266.

Rudberg, M. (2002). Manufacturing Strategy: Linking Competitive Priorities, Decision Categories and Manufacturing Networks, PROFIL 17, Linkoping Institute of Technology, Linkoping, Sweden.

Sherwin, D. (2000). "A Review of Overall Models for Maintenance Management," *Journal of Quality in Maintenance Engineering*, 6 (3), 138-164.

Stoel, M. D. & Muhanna, W. A. (2009). "IT Capabilities and Firm Performance: A Contingency Analysis of the Role of Industry and IT Capability Type," *Information & Management*, 46 (3), 181 - 189.

Taskinen, T. & Smeds, R. (1999). "Measuring Change Project Management in Manufacturing," *International Journal of Operations and Production Management*, 19 (11), 1168 – 1187.

Tuttle, B. & Vandervelde, S. D. (2007). "An Empirical Examination of Cobit as an Internal Control Framework for Information Technology," *International Journal of Accounting Information Systems*, 8 (4), 240-263.

Weil, P. & Ross, J. (2004). 'IT Governance on One Page,' Massachusetts Institute of Technology [Online], [Retrieved November 19, 2009], http://web.mit.edu/cisr/working%20papers /cisrwp349.pdf.

4

Customer Knowledge Management Competencies Role in the CRM Implementation Project

Abdelfatteh Triki and Fekhta Zouaoui

Institut Supérieur de Gestion- Tunis

Abstract

The purpose of this research is to understand the way a CRM project implementation may contribute to the success of the project. The approach of Knowledge Management Strategic Alignment was used as a theoretical framework and a case study was realised for this end. The case study allowed exploring the role of customer knowledge competencies of the company in the CRM project implementation. Three types of customer knowledge competencies are required: customer knowledge acquisition skills, customer knowledge sharing skills and finally customer knowledge use skills.

Keywords: CRM, Knowledge Management Strategic Alignment, customer knowledge management competencies.

Introduction

Given the high rate of CRM project failure that was noticed in the management practice, many researchers tried to identify the key success factors of CRM implementation projects. 50% of managers complain about the failure of their project and an increase of 40 to 75 % in CRM projects is predicted*. The knowledge management capacity of the company (Croteau and Lee, 2003; Chen and Chen, 2004) as well as the strategic alignment (Chen and Chen, 2004) are considered as key success factors in the CRM literature. The central question of this research would be to understand the role that the company strategy of customer knowledge plays in the success of his CRM project. The objective of this paper is to show that the success of a CRM project depends on two elements: 1. The alignment between the Knowledge Management strategy of a company and the CRM strategy. 2. The customer knowledge management company competencies.

Following a review of the CRM and the customer knowledge management concepts, the knowledge management strategic alignment will be proposed. After presenting the methodology used in the case study, we will be discussing and analysing the results that were found.

Theoretical Foundations

Emergence Conditions and Definitions of CRM

Many factors explain the emergence of CRM that is considered as a company strategy oriented towards clients. In a context characterized by commoditization of products, market saturation, increased customer demand and lower loyalty, we are in the context of relationship marketing where the customer conservation becomes a strategic marketing objective.

These changes found in the loyalty strategies were made easier by the development of communication and information technologies. The separation between the firm and the customer noticed in the product view during the 1980's was compensated thanks to the emergence of information systems.

Two stream of research form the theoretical foundation of a CRM concept (Agrebi, 2006); a strategic stream (relationship marketing) and a technological stream related to the information systems. In fact, Crosby and Johnson (2001) identify the customer relationship management as a business strategy that multiplies the use of technology and includes it in all its process to create retention and loyalty over time.

More generally, the focus of the CRM concept is to build a long term and value-added relationship for both business and customers. For this purpose, the company is brought to focus again its efforts and resources on its most profitable customers.

In this perspective, Brown (2001) define the CRM as " a strategy that a company follows to understand, anticipate and manage the current and potential needs of its customers. During this process that involves changes related to strategy, procedures, structures and techniques, a firm struggles to better organize itself around its customers' behavior. This requires the acquisition of knowledge about them and its application at all levels to obtain both profits and customer satisfaction".

This definition highlights the importance of knowledge management in a relational approach. Several studies (such as Zablah and al, 2004) discussed the knowledge

management contribution process in maintaining lasting and profitable relationships between the company and its customers, thus contributing to the success of the CRM implementation project.

In fact, Zablah and al (2004) define the CRM as "a continuous process that results in the use of the market data in order to create and maintain profitable relationships with the customer". These authors focus on the knowledge management process that would make the management of the business-client interaction easier.

In this context, the CRM is considered as skill set as far as the long-term profitable relationship with customers won't be possible unless the companies are able to change their attitude towards the customers individually**. According to Peppers and al (1999) "the CRM means to be able to change the attitude towards the client depending on what he tells us and what we know about him". Besides, the success of CRM relates to the possession of tangible and intangible resources so that the company is flexible to the customers' needs.

Customer Knowledge Management

In the light of the theoretical foundation, a new concept has emerged; it's customer knowledge management. This concept emphasizes the customer knowledge and not the company knowledge about the customer. This was traditionally collected through market research and is no more sufficient to establish innovative concepts within the company (Podslonko and al, 2007).

The focus of this strategic process lies in the active role that the customer plays in the knowledge management strategy of the firm to become real partners of the company in order to make the innovation process of its products and services better (Gibbert and al, 2002).

There are three types of customer knowledge: knowledge for customers (delivered from the company towards the client such as information about the product), knowledge from the customers (their ideas and recommendations concerning the improvement of the product) and finally knowledge about customers (their expectation and needs) (Salomann and al, 2005).

The contribution of the customer knowledge management strategy of a company to its CRM approach is reduced to the fact that the long-term profitable relationship that firms wish to strengthen with clients can not exist unless these companies are adaptable to customer needs. This flexibility depends on tangible and intangible resources held by the company and allows it to adapt its attitude towards the customer individually. "The CRM means to be able and willing to change one's attitude towards a client according to what he says and what one knows about him" (Peppers and al, 1999).

Although the customer knowledge management contribution to the CRM success was widely discussed in literature, all the competencies that firms must have in terms of customer knowledge management and that determine the success of its CRM process have not been identified yet.

Strategic Alignment of Knowledge Management

The research interested in the customer knowledge management did not indicate the organizational mechanisms necessary for their integration into the global knowledge of the company (Garcia-Murillo and Annabi, 2002). The knowledge management strategic alignment model developed by Abou-Zeid (2008) suggests an integration approach of these two types of knowledge allowing identifying the company's competencies in acquiring, analysing and exchanging both organizational knowledge and customer knowledge.

The knowledge management strategic alignment is fundamentally based on the idea that efficient use of a company knowledge is possible only if the knowledge management strategy and the firm strategy are aligned (Abou-Zeid, 2005). Efficient use refers to the gains collected by the company from its Knowledge Management investments. This model consists of four elements: the firm strategy, the organizational infrastructure, the knowledge management strategy and the knowledge management infrastructure. It emphasizes the importance of business skills either on the strategic level, its organizational infrastructure or on the knowledge management process level. These competencies have several dimensions such as technical dimension, organizational ones and finally human dimensions.

According to Abou-Zeid (2008), the necessary and required skills on the knowledge management level are:

• ability to facilitate the exchange and share of knowledge

• ability to develop human and cultural structure in order to promote this exchange

• A predisposition to use the available technologies to create, share and document knowledge

A knowledge management responsible has to combine several abilities: those of specialist in business strategy, an expert in technologies and even those of a professional in human resources (Malhotra, 1997)***

In this research, the objective is to identify specific skills as well as business skills in terms of customer knowledge management during a CRM approach.

Research Methodology

Within this research, a case study was conducted in order to understand the CRM process implementation and identify the business skills and mechanisms in terms of customer knowledge management during the CRM implementation project and the use of this tool. The choice of the case method was motivated by the fact that the study of a CRM project success represents a contemporary phenomenon in a context of real life (Yin, 2003). Two different techniques have been used: the direct observation as well as interviews with call center managers and one of the supervisors.

a) Company presentation: It is a customer service unit of a Tunisian company dealing with the marketing of household products. This call center serves as a mediator between the technical, marketing and commercial service. It ensures two principal missions: the contact with the customer and the transmission of information flows between the client and the relevant departments. Several activities performed within call centers use CRM tools such as complaint management, the management of the activity "insurance" as well as the management of few marketing campaigns (customer registration, evaluation study of the activity "insurance")

b) Research Protocol: the interview guide**** consists of five themes, four of

which correspond to the four components of strategic alignment model and a fifth that is interested in the CRM project evaluation.

Given the criticism targeted to the research based on case studies especially on the level of their reliability, Yin (2003) suggests to create a research protocol that allows reaching reliability: it means conducting the case study in a way that allows other researchers to repeat the same research protocol in order to reach the same results.

This protocol must contain the following elements: the central research question, one or more research propositions, the theoretical framework of the research, the data collection design (including the interview guide already made) and the case study report

c) Analysis and interpretation of results: the results of interviews prove that a company's CRM implementation project success comes from the knowledge management during the integration of the tool and its use. Yet this alignment was not enough, it had to coexist with organizational skills in terms of customer knowledge management

Table 1: Elements of Strategic Alignment in the Case of Company X

CRM strategy of the firm	Stratégie de gestion des connaissances	Infrastructure et Processus Organisationnels	Infrastructure et Processus en matière de KM
-Adaptation to important sectoral changes especially on the level of competition intensification -Market saturation (Household products) and the importance of customer relationship management -Making the call center an important pillar in the CRM business process of the firm (especially on the level of customer satisfaction)	-Data acquisition from clients themselves, from retailers or even other partners	-A small management unit which operates as an outsourcing customer service - Composition: six persons, a supervisor and a manager -Shift from task specialization to customer personalization - Structural independence with the subcontractor company	-Several interactions between the different knowledge management tools within a comapny (CRM, outlook mail, Oracle application...) -Use of procedure manual for the newcomers training -Absence of a KM responsible -Data transmission between call center members in a formal or informal manner -Exploration of CRM data in order to make a reporting -Knowledge transfer through training sessions on products and work procedure -Use of several terminologies to codify knowledge

From this table, the choices of knowledge management strategy support business strategic decisions, that means; the competition intensification and the market saturation compelled the firm to establish knowledge generation process with multiple sources (customers, retailers, and other partners). Besides, the will to make a call center a pillar for a business CRM approach led it to employ several knowledge transfer processes either explicitly (through procedure manuals) or implicitly (in an informal manner). This transfer can also be done through training sessions for new recruits dealing with products, models as

well as different working procedures relative to different activities of customer service.

Finally, this knowledge transfer that seeks to manage quickly and efficiently the customers' claims and thus ensures their satisfaction, is made easier by knowledge coding and the use of specific terminology.

Moreover, the working processes are supported by the company KM tools. One can notice the use of the CRM tool in the customer service, the use of outlook e-mail and other knowledge management tools related to other activities (quotation, installation request...). All these tools aim at facilitating data share and exchange among the call center members on the one hand and between the after sale service staff and the call center staff on the other hand. The organizational processes adopted with the CRM tool are characterized by a shift from a task specialization towards a customization per customer: the operator must deal with the customer request from A to Z. This could be against the CRM approach that consists at centralizing all the customer information in one database, so that everyone can easily access it and thus handle quickly the customer request. This organizational processes change was introduced for more efficiency of the call center and in order to ensure the work performance of customer service. Before installing the CRM technology and even at the beginning of its use, the operators have been specialized by task, some of them deal with customer complaints and others with the insurance requests.

According to the case study results, the CRM strategy alignment with the firm knowledge management strategy is obvious. Its contribution to the project success would be completed by the identification of the principal knowledge management competencies. However, the CRM project success in the call center is claimed by its director, who states that "the project is successful since we have reached functional tool objectives"

Competencies Identification in Terms of Customer Knowledge Management

During the CRM implementation, knowledge was collected within the company headquarters as well as the technical department. When using the software for a complaint management, the call center members must know personal details, data of the products (in terms of purchase date, reference model) as well as information concerning the nature of the complaint (type of failure, the breakdown date.....). The knowledge in this case is accessible according to the task work and transferable in the three company sites through Tunisia.

Customer knowledge acquisition competencies, which are accessible by the staff according to specific rights, embody the first dimension of customer knowledge management skills.

Tacit knowledge of a business staff was converted explicitly through different versions of the user guide. These manuals include procedures to be followed in the customer complaint management, facilitate data sharing among the call center members (especially new recruits). The value system introduced within the call center is based on data sharing between its members, teamwork and collaboration, "very visible concepts in the case of complaints from VIP or in some particular cases where the usual procedure that is followed can not be applied" says the call center supervisor. Abou-Zeid (2005) states that the necessary and required knowledge management competencies are: the ability to facilitate knowledge exchange and its sharing, the capacity to develop cultural and human structure to promote this exchange, the predisposition to use the available technologies to promote creativity, knowledge documentation and sharing.

Transferable knowledge during the CRM project of the call center is the knowledge about the CRM approach, technical and computer related knowledge as well as organizational knowledge (human resources structure and management). The explicit knowledge transfer was performed through CRM project team meetings with the company staff (for the after sale service for instance it concerns people dealing with products fixing). Moreover, the data were coded in order to facilitate its use. Different terminologies are used in the CRM tool such as for the status of the customer record ("pend": under repair", "close: closed form", etc...) or even the intervention type (repair, installation, etc...). Fields were added such as "notes" where the operator takes notes and specific remarks about the intervention (generally communicated by the customers). The implemented CRM tool allows indexing client files in order to describe the file status, thus identifying the task to be performed by the operator according to this file status (for example: when a file is on "close" mode, the operator must perform "a happy call" to evaluate customer satisfaction). Finally, several technological tools have been used equally to the CRM in order to ensure the data exchange between the technical department and customer service (outlook, oracle application, etc...)

Consequently, skills in terms of data sharing are the second dimension of customer knowledge management competencies.

The tacit knowledge of this call center was converted into an explicit form during the CRM implementation through seven different using guide versions including modifications and addition of several functions such as: SMS management aiming at informing the customer that his product is fixed, ensuring file management, installation management, technicians' schedule management and finally management of repair history. These new services have been developed using old knowledge, thus leading to the emergence of new knowledge in the form of new fields that are integrated in the CRM tool (adding notes fields, identifying geographic areas by code). According to Akhavan, Ashar and Heidari (2008), these processes correspond to the knowledge use process that deals with looking for knowledge adapted to solve a problem. These skills may lead to the extraction of a new knowledge that must be recorded for a future use.

The managers' capacity to analyse information recorded in the database (by creating a reporting on the number of repair, the types of failure, the types of fixed products.......) represents another firm capacity of using this acquired knowledge efficiently in order to improve its products and services and maintain long-term relationships with customers.

Thus, the company's skills in terms of using knowledge form the third dimension of customer knowledge management competencies.

The Three Types of Knowledge

During the CRM implementation, knowledge oriented towards the customer, what Salomann and al (2005) call "Knowledge for the customer" is what is most important. Thus, this firm should acquire product list, models, references, types of spare parts as well as work procedures.

In case of using CRM to manage a customer claim for instance, the kind of information that the company should know about its customers are clients' personal information, data related to the product in terms of the purchase date, the reference and information concerning the type of failure. This is what Salomann and al (2005) call "Knowledge about customers". During the use of the CRM tool, this knowledge is acquired by customers themselves. (See Table 2)

Table 2: Types of Knowledge and Customer Knowledge Management Competencies in a CRM Project

	Types of Knowledge	Types of competencies
CRM implementation	Knowledge for the customer	Competencies of acquiring, sharing and using customer knowledge
CRM use	Knowledge about the customer	Competencies of acquiring, sharing and using customer knowledge

Conclusion

The knowledge management approach adopted by this company is an approach based on promoting knowledge sharing between users. Work procedures written in the use guide of the CRM technology facilitates data transfer between the different users. In the case of tacit knowledge, it is about sharing experiences (informal and semi-formal learning). If it is about a new case, the data transfer is often done informally (it is also the case of VIP customers). In the case of explicit knowledge, data transfer is performed through formal learning such as the professional training (Abou-Zeid, 2008). New recruits follow a training about work procedures and the use of CRM tool.

According to the literature and the case study results, the three acquired customer knowledge competencies and which can be identified as dimensions of customer knowledge management are: competencies of acquiring, sharing and using customer knowledge. Nevertheless, this research presents certain limits: we have kept in mind only the theoretical frame of knowledge management strategic alignment to assess the CRM implementation project, while other theoretical fields could have been used such as the change management, the impact of relationships between the tool developers and users. Only one case study isn't enough to understand the customer knowledge management contribution to the CRM success. Several future ways of research might be dealt with: testing the construct pertinence of "customer knowledge management competencies", validate the model on the level of other companies in different sectors.

Acknowledgment

*According to Gartner, quoted by King S.F and Burgess T.F (2008), "Understanding success and failure in customer relationship management, 37,421-431

** Zablah, A.R, Bellenger, D.N, Johnston, W, J (2004), "An evaluation of divergent perspectives on CRM: Towards a common understanding of an emerging phenomenon", Industrial Marketing Management, 15 pages.

*** Quoted by Abou-Zeid (2005), "Alignment of business and knowledge management startegies", M Khosrow-Pour (Ed), Encyclopedia of Information Science and Technology, Vol 1, 98-103, Hershey, PA: Idea publishing Group.

**** See the interview guide in Appendix A and an example of an interview analysis in Appendix B.

References

Abou-Zeid, E. (2008). "Developing Business Aligned Knowledge Management Strategy," in "KM: Concepts, Methodologies, Tools and applications," *Information Science Reference*, Hershey, New York.

Abou-Zeid, E. (2005). 'Alignment of Business and Knowledge Management Strategies,' M Khosrow-Pour (Ed). Encyclopedia of Information Science and Technology, Vol 1, 98-103, Hershey, PA: *Idea Publishing Group*.

Agrebi, M. (2006). "Les Apports, Obstacles et Facteurs clés de Succès d'une e-relation : Le Point de Vue des Fournisseurs de Solutions eCRM," *Communications 5ème journée nantaise de recherche en e-marketing*.

Akhavan, P. & Heidari, S. (2008). "CKM: Where Knowledge and the Customer Meet," *KM Review*, Vol. 11, Issue 3, July-August, p 24-29.

Almotairi, M. (2008). "CRM Success Factor Taxonomy," *European and Mediterranean Conference on Information Systems* 2008 (EMCIS2008). May 25-26, Dubai.

Bennani, A. (2004). 'La Réalité de l'alignement Stratégique et la Formulation Stratégique dans l'entreprise

Pharmaceutique Française, Allemande, Anglaise et Espagnole,' *Actes de l'Association Information Management*.

Benavent, C. (2003). 'CRM, Apprentissage et Contrôle Organisationnel,' Téléchargé du Site: www.christophe.benavent.free.fr

Benavent, C. & Waarden, L.-M. (2001). 'Programmes de Fidélisation : Stratégies et Pratiques,' Les Cahiers de la Recherche CLAREE (*Centre Lillois d'Analyse et de Recherche sur l'Evolution des Entreprises*, IAE Lilles, 33 pages.

Brown, S. (2001). 'CRM: la Gestion de la Relation Client,' *Editions Village Mondial*, 358 pages.

Chen, Q. & Chen, H. M. (2004). "Exploring the Success Factors of eCRM Strategies in Practice," *Journal of Database Marketing & Customer Strategy Management*; 11, 4, pg. 333-343.

Crosby, L. A. & Johnson,S. L. (2001). "Technology: Friend or Foe to Customer Relationships," *Marketing Management*, 10, 4, 10-11

Croteau, A.-M. & Li, P. (2003). "Critical Success factors of CRM technological Initiatives," *Canadian Journal of Administrative Sciences*, 20, 1, 21-34.

Dignan, L. (2002). "Is CRM all it's Cracked up to be," *CNET News.com* (April 3). http://news.com.com/2100-1017-874356.html (l'accès au site a eu lieu le 28 /07/2002).

Fimbel, E. (2007). Alignement Stratégique Synchroniser les Systèmes d'information avec les Trajectoires et Manoeuvres des Entreprises, *Editions Pearson -Village Mondial*

Garcia-Murillo, M. & Annabi, H. (2002). "Customer Knowledge Management," *Journal of the Operational Research Society*, 53, 875-884.

Gibbert, M. & Leibold, M. & Probst, G. (2002). "Five Styles of Customer Knowledge Management and How Smart Companies Put them into Actions ," *European Management Journal* Vol. 20, No. 5, pp. 459–469.

Henderson, J. C. & Venkatraman, N. (1993). "Strategic Alignment: leveraging Information Technology for Transforming Organizations," *IBM Systems Journal*, 32, 1, 4-16.

King S. F. & Burgess T. F. (2008). "Understanding Success and Failure in Customer Relationship Management," *Industrial Marketing Management*, 37, 421–431.

Podsolonko, E., Curbatov, O., Gay, M., Pavlidis, P. Louyot, M.-C. & Bonnemaizon, A. (2007). "La Compétence du Client au Cœur du Customer Empowerment et de la Relation Client," téléchargé du site http://www.nbuv.gov.ua/portal/natural/uzt nu/zapiski/2007/economics/uch_20_1e/pod solonko_34.pdf

Salomann, H., Dous, M., Kolbe, L. & Brenner, W. (2005). "Knowledge Management Capabilities in CRM: Making Knowledge For, From and About Customers Work," Proceedings of the Eleventh Americas Conference on *Information Systems*, pp. 167-178, Omaha, NE, USA.

Yin, K. Robert (2003). 'Case study Research: Design and Methods,' Third edition, *Sage publication*.

Zablah, A. R., Bellenger, D. N. & Johnston, W. J. (2004). 'An Evaluation of Divergent perspectives on CRM: Towards a Common Understanding of an Emerging Phenomenon,' *Industrial Marketing Management*, 15 pages.

Appendix A: The Interview Guide

Theme 1: Presentation of the company, of its products and its general strategy as well as its CRM strategy

1) What are the strategic choices of the company?

2) What are the skills held by the company to be distinguished from the competition?

3) Governance: What are the alliances, the partnerships and the choices in outsourcing performed by the company?

4) Describe the customer relationship management process of the company (Steps and objectives)

Theme 2: ICT strategy of the company (the information technologies in general and the CRM tools in particular)

1) Does the ICT department have a strategic impact within the company?

2) Which style does team management of ICT projects have?

3) What is the integration and complementarity of ICT investments?

Theme 3: ICT infrastructure and processes of the company

1) Describe the data system architecture of the company (applications and technological tools)

2) What are the work processes related to the information system?

3) What are the skills and capacities in terms of ICT? (choices related to staff training and knowledge development of ICT staff)

Theme 4: Organizational infrastructure of the company:

1) Describe the company's organizational structure?

2) Describe the work processes within the company (those related to the product development, the customer service, the marketing, the sales,...)

Theme 5: The evaluation criteria of the CRM project

1) How do you evaluate the success of your CRM project?

2) How has the solution been used? (Frequency and type of use, user profile, the objectives of the use, the type of reporting...)

3) What are the evaluation criteria used to claim that your CRM project is successful)

4) Do you think that after X years of the CRM implementation project, the expected objectives were reached?

Appendix B: Extract of an interview analysis:

Theme/Sub Theme	Interview	Remarks
CRM strategy of the comapny *Strengthen the CRM state of mind in the company (+)* *Direct relationship between the firm and customers (+)* *The firm relationships with retailers and partners (+)*	« The idea of implementing the CRM in the company came after a trip outside the country. It is also related to the CRM state of mind in all our companies. It is our firm itself that was mostly interested in adopting this process, since it has a direct link with the customer's activities (delivery, after sale service,... Since our business calls for direct contact with customers, we can even say that its will is based on this direct contact » « There are privileges granted to important retailers facilitating a delivery for a retailer having a quick sale for instance: he places the order with the commercial and we can make the delivery with our trucks »	*Language :* Using the past tense to describe a past situation *Terminology :* *Use both terms « our company and we » to talk about business thus reflecting a sense of belonging especially that the speaker is part of the family business owner.*
Knowledge management strategy *Technical knowledge from external consultation(+)* *Knowledge combining an adapted price between the software and the consulting services (+)* *A change in the management has led to a change in the tool use (0* *Accessibility to knowledge by all the staff is difficult during the CRM implementation project(-)* **Kowledge Management infrastructure and process :** *Absence of a knowledge management responsible (during the project and currently in the firm (-)* **CRM project evaluation :**	« The choice of the CRM software lasted six months, several suppliers have ben consulted such as Siebel, PeopleSoft in France, but the choice has been focused on Oracle CRM because they have suggested a good financial offer (including the package and the consulting service). « After the suitability study, the CRM implementation has not occured due to a change of management policy in the call center». « It was hard to make the staff understand the reason why the insurance cases would be treated now by the call center. People were wondering if it was because they didn't make a good job. People didn't also understand why procedures have changed... There has been a lack of organization and communication from our part on this level.» « The project team was composed of computer scientists and trades. For instance to integrate the work procedures related to after sale service in the CRM tool, trades belong to the technical company department. The CRM team was headed by a computer scientist » « CRM was not integrated just to earn money, the objective was to make a balance between money and the brand notoriety. » « There has been much progress in the CRM : philosophy and interest in CRM are created however the level of contact personalization has not been reached yet (to be "friends" with the customer) » « The CRM satisfies the customer but it has nothing to do with the service quality »	*Language :* Using the past tense to describe a past situation *Terminology :* *Use of the passive form to describe a brief event (change) that occurred in the company.*

5

A Framework for the Construction of Ontology for ICT Experts

Akmal Aris, Juhana Salim, Shahrul Azman Mohd Noah and Kamsuriah Ahmad

Universiti Kebangsaan Malaysia, Bangi, Malaysia

Abstract

Systems that help to find answers to suitable experts have received the attention of many researchers. Pioneer researchers on the development of search system for experts emphasized the importance of a search system that required returning a list prioritizing the names of individuals. Among the issues raised related to locating experts are, the critical problem of maintaining up-to-date information in the expert database and the inadequacy in expert finding systems in returning search results that are expected to account for not only the list of the names of the experts, but any information related to the experts and those involved with them. In the last decade, researchers have examined the search for experts from various research perspectives such as expert tracking system and construction of expert profiles involving ontology. This paper aims to describe a new approach in designing a framework for the construction of ontology that will be used in the directory of ICT experts. The researchers propose to incorporate thesaurus in their construction of ontology on ICT experts by providing a profile of the experts including their social profile and whatever concerns that may be associated with the expert. The researchers constructed the ontology on ICT experts by extracting information from sources such as their resumes and personal web sites to obtain the index glossary of words that characterized experts to be used in the directory of experts system. In this study, the standards contained in the Performance Evaluation System of the National University of Malaysia are used to extract information on the academicians in the Faculty of Information Science and Technology. The researchers employed the index glossary, metadata of experts and integrate appropriate taxonomy, thesaurus and classification schemes; such as Association for Computing Machinery (ACM) taxonomy, web classification schemes such as Standard Industrial Classification (SIC), North American Industry Classification System (NAICS) to enrich the ontology of ICT expertise. The proposed framework aims at helping users find information on the expert they require and at the same time obtain other information related to the experts from various perspectives encompassing research, consultancy, links with research partners, other interest related to a particular field and other resources. With a semantically driven directory of ICT experts, matters related to ICT can be referred to the right experts. The framework will be validated by developing the directory of ICT experts prototype and by involving domain experts to evaluate the content of the ontology constructed. In addition, the researchers will evaluate the search results of users who will use the prototype to search for ICT experts.

Keywords: Expertise Tracking, Expertise Ontology, Profiles of Experts, Access to Experts.

Introduction

Experts are individuals who play an important role in the success of an organization and are considered critical in creating a value to the organization. The expert skills and knowledge very often can be channeled through consultation, mentoring systems and corporate memory. According to D'Amore (2008), experts are generally located in the formal or informal workspace and their work is based on the domain and culture of the organization. Yiman-Seid and Kobsa (2003) have identified several goals in allocating the experts based on defining the problem, evaluation and analysis, filtering information and project assignments. However, there are problems in the placement of experts in an area where there are difficulties in finding them based on the work context and work locations that are not parallel (Yiman-Seid and Kobsa 2003). Studies on the search for experts have focused on the task of seeking individuals who have the skills and knowledge that are particularly suited to answer the question "Who is an expert in the field of X?" The task to find answers to suitable experts have received the attention of many researchers and most of them concentrated on experts profiling with search applications using query that describe the areas of expertise being sought. Pioneer researchers on the development of search system for experts emphasized the search system that required to return a list that prioritized the names of individuals (Balog and de Rijke 2007, Hawking 2004, Karimzadehgan, White and Richardson 2009, Stankovic et al. 2010). Among the issues raised in relation to locating experts are, the critical problem of maintaining up-to-date information in the expert database and the inadequacy in expert finding systems in returning search results that are expected to account for not only the list of the names of the experts, but any information related to the experts and those involved with them. For example, in tracking an expert, the details that one would be interested in, include: the area of expertise, who are working with the expert, other information needed to contact the expert and the individual / organization or others that have the same expertise and relationships with the expert. Therefore, to develop a system that allows the tracking of experts, there is an imperative need to identify the complete metadata in profiling an expert. These include describing the collaborative environment of the expert.

Initial approach used to search for experts involved the development of a database that stored information about the skills and knowledge of individuals in an organization (Maron et al. 1996; Davenport and Prusak 1998), and merely focused on how to integrate the database containing the same information found in a data warehouse where it can be mined to obtain information on the expert. Most of the earlier researches were carried out by communities in the area of knowledge management_ and their research output includes yellow pages and systems for searching experts (Becerra-Fernandez 2000).

In relation to the research on the profiling of experts to support the development of the directory of experts, another approach used by previous researchers is the use of ontology (Stankovic et al. 2010). Potential use of ontology in data related to the experts has been demonstrated in studies by Meza et al. (2007) where the researchers have combined Resource Description Framework (RDF) vocabulary to search for experts. Meza et al. (2007) designed a framework for the reuse and expansion of vocabulary / thesaurus in the semantic web. The framework is designed to support efforts to develop the ontology considered important in allowing the expert tracking system to find information about an expert not only in terms of qualifications and brief resume, but whatever concerns that may be associated with the expert. As long as such information exist on the Internet, the system can find it. To date, literature on expert tracking system indicates that very little research has been done in the building of ontology to support expert tracking.

This study will develop an ontology-based approach applied in previous studies that used thesaurus and ontology as a method to extend the search for information in various domains. The researchers propose to incorporate thesaurus in their construction of ontology on ICT experts by providing a profile of the experts including social profile where other things and other parties related to the expert could be found.

The body of this paper is organized as follows: Section 2 discusses the literature review; Section 3 defines the research objective followed by Section 4 which describes the research method. In Section 5, the researchers offer their conclusion regarding the potential outcome of their framework and the feasibility of their approach in tracking experts and linking other aspects related to the expert.

Literature Review

Studies in search of experts focused on the task of seeking individuals with appropriate skills and knowledge. Since the last decade, researchers have examined the search for experts from various research perspectives such as expert search system (Hawking 2004; Fu et al. 2006; Balog and de Rijke 2007; Chen et al. 2007;Haruechaiyasak 2009; Karimzadehgan, White and Richardson 2009; Tu et al.2010), construction of expert profiles (Whittaker et al. 1997; Ackerman and Halverson 1998; Crasswell et al. 2001; Krulwich and Burkey 1995; Mockus and Herbsleb 2002; Yiman-Seid and Kobsa 2003; Trajkova and Gauch 2004; Liu et al. 2005; Fu et al. 2006; Yang et al. 2008 ; Serdyukov 2009; Reichling and Wulf 2009; Stankovic et al. 2010), identification of expert (Hinds 1999; McDonald 2001; Pipek et al. 2003; Zhang et al. 2007) and ontology. This literature review has critically evaluated expert tracking system, expert profile and ontology on expert.

Expert Tracking System

Development in research on expert tracking system shows that it is a fast-growing field and the issues discussed above have been given much attention especially on the approaches used to address the related issues and problems. Previous researchers focused on the task of creating a complete profile of experts and integrate ontology development to be applied with semantic web technology. Research effort in profiling experts is a new endeavour.

Studies on the profiling of experts in the expertise search systems have attracted the attention of many researchers. Previous researchers have examined the profile of experts from various angles such as research in the construction of expert profiles (Whittaker et al. 1997; Ackerman and Halverson 1998; Crasswell et al. 2001; Mockus and Herbsleb 2002; Yimam-Seid et al 2003; Trajkova dan Gauch 2004; Liu et al. 2005; Fu et al. 2006; Serdyukov 2009; and Reichling and Wulf 2009) and the identification of expert and use of research information (Wu et al. 2010; Latif et al. 2010). Expert search approach begins with the search by using profile information to obtain results matching the experts (Crasswell et al. 2001; Liu et al 2005). The search for expertise based on profile is the first step in automating the search for expertise in the organization and to avoid manual maintenance of personal profile information such as resume and personal web pages (Serdyukov 2009). According to Trajkova and Gauch (2004), profiles can be constructed based on questions raised or based on the observation of the user activity. A user profile is usually delivered using keywords or vector concepts.

In particular, researchers have focused on topical and social profiling of experts. Yang et al.

(2008) highlighted some of the questions to be answered in building the profile of expert which includes: "What is the area specialization of the expert?", "Who is the expert in a particular area?", "Who is familiar with this type of expertise?", "With who is the expert working?", "What is the detailed information to enable the experts to be contacted?" and "Who are the individuals / organizations or parties that have the same expertise and relationships related to the experts?"

To answer the above questions, previous studies have developed systems that can assess the expertise in individuals and determine the expert best suited with whatever issues that need to be addressed and problems to be solved. Mc Donald (2001) points out that there are problems faced in terms of matching the appropriate documents and expert on the problems faced. He proposed several changes to be made in order to match the human expert with issues and problems to be solved. Mc Donald proposed that humans can assess the expertise in individuals and systems can determine the appropriate expert based on the performance demonstrated in solving a particular problem and the level of performance measurement of the expert to determine the appropriate experts with topics searched. Zhang et al. (2007) highlighted two steps in the search for experts in the social network; that is, by using personal information to determine the scores and the propagation-based approach to identify existing relationships with the experts. Li et al. (2006) look at personal contact information from the perspective of social networks that can be connected through four types of relationship: knows, collaborates, collaborated and consulted by.

Balog et al. (2006) proposed two strategies in the search for experts. First, directly model an expert's knowledge based on the documents that they are associated with and second, locate documents on the topic and then find the expert associated with the topic. According to Krulwich and Burkey

(1995), the involvement of individuals with forums discussing particular topic on the Internet can help in building expert profile. Haruechaiyasak et al. (2009) stressed on the fact that the main purpose of searching for an expert is to identify persons who have specialized knowledge. As expert profiling refers to the study that focuses on the identification of areas of expertise with a specialist, it is important to note that expert profiling needs to be explored in the research involving experts searching.

Previous researchers have developed a number of tools that can help in the search for experts. Fu et al. (2006) have designed the description document model known as 'Person Description Document' for a more effective search for experts. Information such as the features and context of the relevant experts are extracted to build an expert profile known as self description documents for experts. However, several issues were discovered relating to the information resources that are of various types such as websites, electronic mail and databases. In addition, documents containing information of different experts that are found in different formats such as in HTML, PDF and WORD, complicate the extraction process and the consolidation of useful information. Fu et al. (2006) found two major problems encountered in searching experts in the organizations. First, the information on experts are distributed across organizations from different sources and in different formats. Second, most of the information on experts is not fully documented and only part of the document contains information on the experts involved. To overcome these problems, Fu et al. (2006) used Person Description Document (PDD) to compile expert information to a central unit that can be used to profile experts. The advantage of using the self description document is to enable the extraction of expert information to be more flexible and easier to read. In other words, self description document is the index information about the expertise that allows efficient access to the documents related to the experts.

Reichling and Wulf (2009) proposed the application of expert search incorporating two mechanisms in the construction of the expert profile that are: the generation of keywords semi automatically and the use of Yellow Pages approach. Through the construction of keyword profiles, a list of large-scale keywords acquired from the doc., pdf., html and txt file selected from the file system users can be generated. In building the right profile and to protect the privacy issue, experts are allowed to choose their own files or folders containing their documents during the generation of keywords. To obtain a more accurate keywords profile, there are words such as stop words that need to be filtered and disposed. The second mechanism proposed by Reichling and Wulf (2009) in the construction of expert profiles is to use the Yellow Pages approach where experts will enter the contact information and other information, such as educational background, job description, qualifications, competency of other language and other personal information. According to Reichling and Wulf (2009), using the Yellow Pages approach, experts will develop a profile of their expertise.

Ontology on Expertise

Another subject of concern in relation to some previous researches in expert tracking system is ontology. Ontological approach in the search for expert is to identify appropriate expertise to the keywords found with additional information related to the results given by the expert search system. Among the researchers who have studied the use of ontology in the expert tracking system are Liu et al. (2007); Meza et al. (2007); Stankovic et al. (2010) and Punnarut and Sriharee (2010).

Liu et al. (2007) proposed the use of ontology on experts to integrate the various indicators of expertise from diverse data sources and the use of domain ontology to replace the search for experts based on concept rather than on keyword. The purpose of Liu's study

is to increase the chances of finding the relevant experts and assist users to select the appropriate expertise by providing more detail information for each expert. According to Liu et al., domain ontology is constructed not only to save the key concepts but also the concepts related to a specific domain found in a collection of documents.

Xing et al. (2009) whose research incorporated the use of thesaurus on ontology construction explained that there are four key elements in ontology development; they are: terms, hierarchies, semantic network and ability reasoning. According to Chang and Lu (2001), hierarchy is another important element of ontology. Hierarchy among different object classes refers to inheriting relationship (is-a, kind-of, part-of), while hierarchy among different classes refers to combination relationship (intersection, union, inverse set, complementary set of other classes). By this way, terms (concepts) can be connected together by hierarchy. The fact that thesaurus as vocabulary table also has hierarchy makes it, after a little change and process, possible to be used in ontology. Xing et al. (2009) justified the reason why thesaurus is useful in building ontology as follows: (i) the standard of terminologies and professional division of thesaurus can satisfy the request of clarity, the completeness and coherence on ontology; (ii) the extendibility of terms so that one can continue adding new terms without changing original terms; (iii) to save time in building ontology because it consumes a lot of time if we fully depend on domain expert. However, the domain expert is still needed to add more attributes and relationships to the ontology because thesaurus is found to be lacking of relationship (Lauser et al. 2006 and Xing et al. 2009). Thus, thesaurus is necessary in building efficient ontology on expertise.

Such approach is to overcome the weaknesses faced in the traditional way of finding experts that commonly requires database of expertise/skills. Punnarut and Sriharee (2010) used data mining techniques to identify the expertise of researchers with skill classification ontology. Skill classification ontology is a model of expertise that contains the expertise in the field of computer and information science. The purpose of Punnarut and Sriharee using skill classification ontology and research profile of experts is to develop an expertise search system with semantic matching.

In this context, the present study will identify the information related to the search for experts on ICT based on previous research studies resources. Based on the analysis of information search on experts, the researchers found that previous researches have focused on semantic search of experts using the names of the experts. There is also a research effort to search for experts based on the experts' publications searchable from the database containing articles and papers of experts. But for a good expertise search system, the system not only need to look for the expert that need to be tracked to solve issues , but should be able to make relationships with the experts from various perspectives such as research, consultancy, links with research partners, the interest in a particular field and other resources related to the experts. In the context of this study, this study will not be doing the identification of the status of the expert or measure the levels of expertise, but will focus on providing results that match experts with the information required by the user. Different perspectives need to be taken into account to enrich the relationship which exists in connection with experts to get results that are relevant to a search topic. In the context of this study, sources of information on the researchers in the research groups of the Faculty of Information Science and Technology, National University of Malaysia will be used. All the information available to the researcher's resume will be extracted and stored in the word index to be used in the construction of ontology of ICT experts.

Research Objective

In order to fill in the research gaps, the main objective of this research is to design a framework for the construction of ontology on ICT experts based on expert profile. To achieve the main objective, the following sub-objectives were identified: i) to explore the approaches in profiling experts as the basis to the construction of ontology on ICT experts; ii) to integrate thesaurus and classification systems in the development of a semantically driven directory of experts; and iii) to evaluate the effectiveness of the expert directory.

Methodology

To achieve our first research objective, the researchers of the present study conducted literature search from various databases such as ACM, Science Direct, Springer Link and Citeseer to acquire research based articles on expert tracking and the profiling of experts. They then analyzed the approaches that past researchers used in solving the problems related to tracking experts and the profiling of experts. In relation to past researches on the profiling of experts, it is discovered that there were researchers that used ontology. Specifically, it was found that researchers used thesaurus and classification schemes in constructing ontology. Therefore, to achieve our second research objective, an expertise ontology will be constructed which will be explained in more detail in our framework.

Based on the critical evaluation of past research and literature related to the areas mentioned earlier, a framework was designed for the construction of ontology on ICT expertise based on expert profile (See figure 2). As shown in the framework, the area that will be studied includes:

I.   Experts system and Profiling of experts.

II. Semantic Web Technology.

III. Ontology on ICT expertise.

IV. Thesaurus, taxonomy and classification scheme.

V.   Directory of experts.

The framework begins with the identification of major problems faced by users in finding experts and expertise based on user demand. Semantic web technologies will be studied in order to apply in the construction of ontology on ICT experts to provide more relevant search results for expert tracking. The information needed will be extracted from various sources such as resume and personal website to be used in the construction of ontology on ICT experts.

To achieve our second objective, based on previous studies by Fu et al. (2006) who used Person Description Document (PDD), the ontology on ICT experts will be constructed by extracting information from sources such as resume and personal web sites to obtain the index glossary of words that characterized experts to be used in the directory of experts system. In this study, the researchers of this paper will use the standards contained in the Performance Evaluation System of the National University of Malaysia to extract information on the academicians which include the following information: personal, education, teaching, supervision and reviewing, publications, research, consultancy, conferences, invention / innovation / product, or contributions to the university administration, community services / activities, student services, information services, professional qualifications, membership of professional bodies, awards and insignia and training / short courses / workshops [See Figure 1].

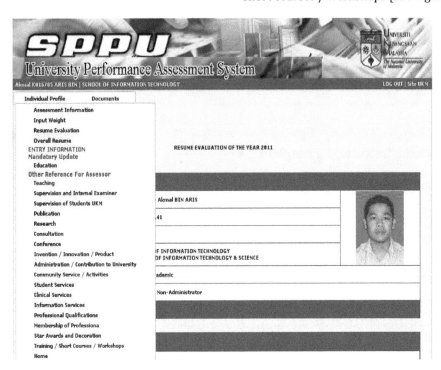

Figure 1. Information Available on University Performance Assessment System

By having this ontology of ICT experts, the expert directory will not only be used to help accelerate the search but also show the relationship between experts with other areas such as publication, research, conferences and other activities and other people associated to the experts. The researchers will use the index glossary, metadata of experts and integrate appropriate taxonomy, thesaurus and

classification schemes, such as Association for Computing Machinery (ACM) taxonomy, web classification such as Standard Industrial Classification (SIC), North American Industry Classification System (NAICS) to enrich the ontology of ICT expertise. The ontology of ICT expertise is the main engine for semantic web retrieval which is a technology for semantic web.

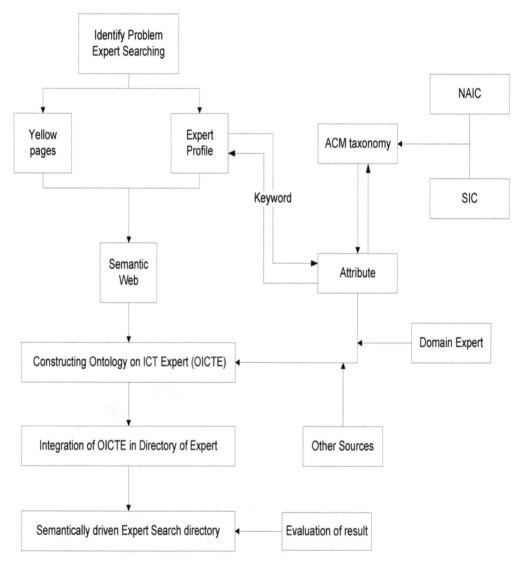

Figure 2. Ontology Development Framework for the Directory of Experts

Once the ontology on ICT experts had been constructed, a prototype directory of expert is then developed as a source of reference to facilitate the tracking of experts and their areas of specialization. It will be developed using System Development Life Cycle (SDLC) in order to validate our framework. The ontology constructed will be integrated into the directory of experts to enable the directory to provide a semantically driven search capability. The semantically driven search capability will be tested by matching the search result with the requirements of users that used the directory of experts.

There are two validation processes involved in order to validate our framework. Firstly, the ontology of ICT experts will be validated by the experts in the ICT domain to evaluate the concept, attribute and relationship which existed in the ICT experts ontology that have been constructed. The second validation process will involve content evaluation. The users will be instructed to use the ICT expert directory prototype to search an expert based on their query. The result from the query term which is entered by the user in the directory will be tested to know whether the results are relevant and meet the requirements of users. A qualitative study will be conducted to obtain feedback from users about the effective use of ICT experts ontology in   directory of ICT experts prototype.

Conclusion

An important outcome from this research is a framework for building ontology of ICT experts for an ICT expert directory which applies a new approach in creating ontology. This research involves the application of a new approach to create ontology by using well established thesaurus such as Library Congress Subject Heading and ACM Taxonomy including web classification schemes such as NAIC and SIC and various references sources. This proposed framework aims at helping users find information on the expert they required and

at the same time obtain other information related to the experts from various perspectives encompassing research, consultancy, links with research partners, other interest related to a particular field and other resources. With a semantically driven directory of ICT experts, matters related to ICT can be referred to the right experts.

References

Ackerman, M. S. & Halverson, C. (1998). "Considering an Organization's Memory," *Proceedings of th0e 1998 ACM Conference on Computer Supported Cooperative Work (CSCW '98)*, ISBN: 1-58113-009-0, New York, NY, USA, 39-48.

Aleman-Meza, B., Hakimpour, F., Arpinar I. B. & Sheth A. P. (2007). "SwetoDblp Ontology of Computer Science Publications," *Web Semantics: Science, Services and Agents on the World Wide Web*, 5 (3), 151-155.

Balog, K. & de Rijke, M. (2007). "Determining Expert Profiles (with an Application to Expert Finding)," *Proceedings of the 20th International Joint Conference on Artifical Intelligence (IJCAI'07)*, San Francisco, CA, USA, 2657-2662.

Becerra-Fernandez, I. (2000). "Facilitating the Online Search of Experts at NASA Using Expert Seeker People-Finder," *Proceedings of the Third Int. Conf. on Practical Aspects of Knowledge Management*, Basel, Switzerland, 30-31.

Chang, C. & Lu, W. (2004).'From Agricultural Thesaurus To Ontology,' *5th AOS Workshop*. 1-4.

Craswell, N., Hawking, D. , Vercoustre, A.- M. & Wilkins, P. (2001). "P@noptic Expert: Searching for Eexperts Not Just for Documents," Ausweb, [Online], [Retrieved October   22,2011], http://es.csiro.au/pubs/craswell_ausweb01.pdf

D'Amore, R. J. (2008). "Expert Finding in Disparate Environments," Thesis (PhD), *University of Sheffield*, Sheffield, United Kingdom.

Davenport, T. H. & Prusak, L. (1998) Working Knowledge: How Organizations Manage What They Know, *Harvard Business School Press*, Boston.

Fu, Y., Xiang, R., Zhang, M., Liu, Y. & Ma, S. (2006). "A PDD-Based Searching Approach for Expert Finding in Intranet Information Management," AIRS 2006, 43-53.

Haruechaiyasak, C., Kongthon, A. & Thaiprayoon, S. (2009). "Building a Thailand Researcher Network Based on a Bibliographic Database," Proceedings of the 9th ACM/IEEE-CS Joint Conference on Digital libraries, ISBN: 978-1-60558-322-8, New York, NY, USA, 391-392.

Hawking, D. (2004). "Challenges in Enterprise Search," Proceedings of the 15th Australasian Database Conference, Darlinghurst, Australia, 15-24.

Hinds, P. J. (1999). "The Curse of Expertise: The Effects of Expertise and Debiasing Methods on Predictions of Novice Performance," *Journal of Experimental Psychology: Applied*, 5, 205-221.

Karimzadehgan, M., White, R. W. & Richardson, M. (2009). "Enhancing Expert Finding Using Organizational Hierarchies," Proceedings of the 31th European Conference on IR Research on Advances in Information Retrieval (ECIR '09), ISBN: 978-3-642-00957-0, Toulouse, France, 177-188.

Krulwich B. & Burkey C. (1995). "ContactFinder: Extracting Indications of Expertise and Answering Questions with Referrals," *Fall Symposium on Intelligent Knowledge Navigation and Retrieval, Technical Report FS-95-03*, The AAAI Press, 85- 9.

Latif, A., Afzal, M. T. & Tochtermann, K. (2010). "Constructing Experts Profiles from Linked Open Data," 6th International Conference on Emerging Technology (ICET), ISBN: 978-1-4244-8057-9, Islamabad, 33- 38.

Liang, A. C., Lauser, B., Sini, M., Keizer, J. & Katz, S. (2006). "From AGROVOC to the Agricultural Ontology Service/Concept Server, An OWL Model for Managing Ontologies in the Agricultural Domain," Proceedings of the 2006 International Conference on Dublin Core and Metadata Applications: Metadata for Knowledge and Learning, 68-77.

Li, J., Boley, H. Bhavsar, V. C. & Mei, J. (2006). "Expert Finding for eCollaboration Using FOAF with Rule ML Rules," Proceedings of the 2006 Conference on E-Technologies, Montreal, QC, Canada, 53-65.

Liu, P., Liu, K. & Liu, J. (2007). "Ontology-Based Expertise Matching System within Academia," *Wireless Communications, Networking and Mobile Computing*, ISBN: 978-1-4244-1311-9, Shanghai, China, 5431-5434.

Liu, X., Croft, W. B. & Koll, M. (2005)."Finding Experts in Community-Based Question-Answering Services," *Proceedings of ACM CIKM*, New York, NY, USA, 315-316.

Maron, M. E., Curry, S. & Thompson, P. (1986). "An Inductive Search System: Theory, Design and Implementation," *IEEE Transaction on Systems, Man and Cybernetics*, vol. SMC-16, No. 1, 21-28.

McDonald, D. W. (2001). "Evaluating Expertise Recommendations," Proceedings of the 2001 International ACM SIGGROUP Conference on Supporting Group Work (GROUP '01), New York, NY, USA, 214-223.

Mockus, A. & Herbsleb, J. D. (2002). "Expertise Browser: A Quantitative Approach to Identifying Expertise," Proceedings of the 24th International Conference on Software Engineering (ICSE '02), ISBN:1-58113-472-X, New York, USA, 503-512.

Pipek, V., Hinrichs, J. & Wulf, V. (2003). Sharing Expertise: Challenges for Technical Support, *MIT Press*, Cambridge, 111-136.

Punnarut, P. & Sriharee, G. (2010). "A Researcher Expertise Search System Using Ontology-Based Data Mining," Proceedings of the Seventh Asia-Pacific Conference on Conceptual Modelling - Volume 110 (APCCM '10), ISBN: 978-1-920682-92-7, Brisbane, Australia, 71-78.

Reichling, T. & Wulf, V. (2009). "Expert Recommender Systems in Practice: Evaluating Semi-Automatic Profile Generation," Proceedings of the 27th International Conference on Human factors in Computing Systems (CHI '09), ISBN: 978-1-60558-246-7, Boston, MA, USA, 59-68.

Rui, S., Qin, T., Shi, D., Lin, H. & Yang, Z. (2007). "DUTIR at TREC 2007," *Blog Track*, *http://trec.nist.gov/pubs/trec16/papers/dalianu.blog.final.pdf*

Serdyukov, P. (2009). Search for Expertise going Beyond Direct Evidence, Thesis, (PhD). *University of Twente, Volgograd*, Russia.

Stankovic, M. Jovanovic, J. & Laublet, P. (2010).'Enhancing Linked Data Metrics for Flexible Expert Search on the Open Web,' 108-123, *Springer-Verlag*, Berlin.

Trajkova, J. & Gauch, S. (2004). "Improving Ontology-Based User Profiles," *Proceedings of RIAO, University of Avignon (Vaucluse)*, France, 380-390.

Tu, Y., Johri, N., Roth, D. & Hockenmaier, J. (2010). "Citation Author Topic Model in Expert Search," Proceedings of the 23rd International Conference on Computational Linguistics: Posters, Beijing, China, 1265-1273.

Whittaker, S., Issacs, E. & O'Day, V. (1997). "Widening the Net Workshop Report on the Theory and Practice of Physical and Network Communities," *SIGCHI Bulletin*, 29(3), 27-30.
Xin, X., Ru, L. & KaiYing, L. (2009). "Building Ontology Base on Thesaurus," 2nd International Conference on Biomedical Engineering and Informatics, ISBN: 978-1-4244-4132-7, 1-4.

Yang, K.- H., Chen C.- Y., Lee, H.- M., & Ho J.- M. (2008). "EFS: Expert Finding System Based on Wikipedia link Pattern Analysis," *Systems, Man and Cybernetics*, ISBN: 978-1-4244-2383-5, 631-635.

Yimam-Seid, D. & Kobsa, A. (2003). "Expert Finding Systems for Organizations: Problem and Domain Analysis and the DEMOIR Approach," *Journal of Organizational Computing and Electronic Commerce*, 131.

Zhang, J., Ackerman, M. S. & Adamic, L. (2007). "Expertise Networks in Online Communities: Structure and Algorithms," Proceedings of the 16th International Conference on World Wide Web (WWW '07), ISBN: 978-1-59593-654-7, Banff, Alberta, Canada, 221-230.

Zhang, J., Tang, J. & Li, J. (2007). "Expert Finding in A Social Network," 1066-1069, Springer-Verlag Berlin, Heidelberg.

6

Measuring Knowledge Management: A Strategic Approach to Library Information Services

Rosângela Formentini Caldas

São Paulo State University, Information Science Department, São Paulo, Brazil

Abstract

In the study of theoretical trends in Administration, the management of information follows the development of theories of Administration; constant-adaptations are suffered. Information Science area understands and concerns itself with the changes wrought in these endeavour of the knowledge society as new forms of communication and integration. The libraries interact in ways that maximize access to information and facilitate the improvement on their structural environment as strategic approach for your services. The research aims at identifying the requirements and specifications of an information system for knowledge management in the public's library environment and proposes to achieve a pre-defined structure for the implementation of administration management. The research is conducted with public's library of the metropolitan region of the North's Portugal. Portugal libraries are institutions that operate in the social process of their communities reflecting the society and its organic sphere of informational performance. These libraries have developed the organizational theories to make a framework easily for effective management practices and have been using their produced knowledge in the optimization of their actions. In the improvement of systems, theoretical administrative trends become management decisions and result in the ultimate success of the organization. In order to achieve its objectives the study verified an economy based on knowledge management, and its production emphasizes the human capital that permeates the condition of the information in support for the development of communities.

Keywords: Organizational Theory; Knowledge Management; Public's library.

Introduction

When the organizations think in their future, they think about improvement process with the perfect symmetry of innovation and order to the very common of its operation. The relationship itself geared input services and employees or products and consuming, they become an organizational concern in how to do your products; how to do the best way to the market and how to make your goals possible.

The construction of a connection among administrative theory streams, and the management process of an organization, aims at formalizing the necessary knowledge for the growth of their structural and environmental frames.

The research aims at developing an organizational model based on knowledge and contributing for the formulation of a theory on the information for libraries, within acquired knowledge and administration experience of human resources.

The knowledge is appreciated, not only as a value itself, but also as power of wealth generator. However, to apply the knowledge in the organization involves intelligence and learning. Intelligence and learning are understood as faculties of thinking and innovation of the processes.

In order to understand the analysis of knowledge the steps demonstrate an illustration sampling has been developed for the learning organization. Another step was necessary data collect to find the creation of a model organization specifically for library. The scientific methodology was to do activities using qualitative research methodology for Portugal's Library.

And the work research analysis involves collecting information for the organization using several methods: observation in technical visits, interviewing and other more specialized job analysis methods such as position or functional analysis. Organizations sometimes use a combination of job analysis methods.

With the continuous analysis in the field of administrative theory for libraries and the social progression impact for the organizational environment is possible to understand the need of valuing generation programs for the knowledge management in these areas. Effective help is mportant to create job descriptions and performance standards that are useful tools for development and performance knowledge management.

In future researches, could demonstrate that universe of organizational knowledge by means of statistical analysis, and evaluation of the data collect for the models in the European Union. Today these models are concept and practical net for the urban knowledge management of the human capital.

Development of Management Thought

In a research study by Chiavenato (2000) one organization is a community of persons with one goal. One organization is a better union for: Human resource; structure; environment; management processes; information. Organizations are social construction of knowledge for the organization's goals.

The theoretical movements look forward to a reflection on the way in which information is processed and offer support to the functioning of the organizations through their historical process. The administration process involves people and resources in the search for the realization of organizational objectives and when conscientiously carried out – according to the spheres inserted in this ambit – develops the variety of activities implied while: planning, organizing, leading and controlling. Traditional functions in this process may be improved with the influence of the experience of the organizations' employees, structuring self-confidence and the connection amidst the speciality of the operation developed by the individuals.

With the development of the way of thinking of the administrative theories, the quality improvement of the offer of a good and/or product and the best management practises in the administration appeared. There isn't a best unique formula to reach the organizations' highly varied objectives, since their environment becomes highly varied as well.

According to Bateman and Snell (1998), the administrative approaches developed according to the needs of confronting the administrators and their modifications along the years while attempting to explain real questions faced by them with the objective of providing tools for the resolution of future problems.

In the applicability of the administrative functions, the theory strengthened the belief of management supremacy in the organizations. It didn't only refer to producing in a great way, like the Scientific Management School proposal, but managing the rising production, which took place after the Industrial Revolution.

The relations of interaction between organizations and environments started to explain more deeply, aspects of the organizational structure and operational processes used in companies. For example, the Systems Theory defended the importance of inter-relations among parts of an organization with its surroundings. The result was that the study of internal factors of an organization weren't enough to help managers in decision making.

The evaluative condensing of the administrative thought appeared through the construction of the steps of different theoretical movements. The results converged since diverse branches of knowledge, were able to be explained well when they started considering their study objectives as systems. We can see, therefore, the growth of the more contemporary organizational administrative management, from the use of topics exposed by the Systems Theory.

It is possible to notice that the evolution of the administrative theories accompanied with the civil movements' history and through this course the developments of the organization were defined. Thus, in the 60s we could already observe attitudes in the human conception of the organizational values as a reflection of the great dissatisfaction of the social revolutions witnessed. In the 80s, Japan reappears as an example of culture to be followed by world economy, culture of competitive strategy which the rest of the nations describe the epistemology of the word with its organizational cultures. In the 90s, subsequently, is adopted the managerial focus of culture while strategy of survival and change in the political, social and ethical sphere and leadership as a promotion of this focus.

Modern and contemporary society, lies on the organizations: it is basically a society of organizations and its different ways of evaluating objectives shows that the managers should consider factors of constant change originated from the new administrative management attitudes, in its decision making (Kaplan and Norton, 2000).

One of the great changes seen in the construction of the administrative theories is the creation of values that cannot be measured anymore. The value of information, knowledge, is the capital that holds the current table of the administrative theories. Current society and its organizations possess new communication channels, new ways of work and interaction which change the structure of a former power, to the power of knowledge. Capital society now becomes classified and described as the society of knowledge (Figure 1).

In the information based society, the resource that stands out the most to an organization is the intellectual capital based on knowledge. Knowledge may be understood as information which is structured, contextualized and full of content to who detains it. Costa, Krucken and Abreu (2000, p. 31) considers that knowledge is a cognitive process, which needs information as raw-material to let it happen.

Figure 1. Relation of the Organizational Theory in the Organization Structure

Another question pointed out by Drucker (1994) in this society characterized by knowledge, is the capacity of decision making, based on the proximity to performance, market, technology and environment, all should be seen and used as opportunities for innovation, all should understand what each one is saying. Technology is one of the components of Knowledge Management. Thus, human talent, supported by computational technology and communication could be considered the competitive differential among organizations.

Methodology

The European Union, worried about constant prosperity, possesses guidelines to care for stability maintenance and promotion of certain social spheres, and, therefore, has certain responsibilities in what concerns its stability. This follows the decisions made by the heads of government and their delegations in a strict control of the process. In the strategy of sustainable development, a concept of assuring a high level of social cohesion and economic prosperity was created.

The issue of generalization has appeared in the literature with regularity. It is a frequent criticism of case study research that the results are not widely applicable in real life. Yin in particular refuted that criticism by presenting a well constructed explanation of the difference between analytic generalization and statistical generalization: In analytic generalization, previously developed theory is used as a template against which to compare the empirical results of the case study (Yin, 1989).

The collect data was made in Guimarães City, Portugal. In the General Report on the activities of the European Union (2006), this city is the next step for the Europe model in net of the practice management urban knowledge. Some cities models are examples as well as in France or Scotland. The local for the interview and technique visit, was the Raul Brandão Library (Figure 2).

Figure 2. Raul Brandão Library

The theory was formulated before the beginning of the collection of data, and consequently, proposed to aid the questions in the chart composition. The data arrangement was available in five phases: determination of the collecting techniques and registration of data; collection of data; establishment of data analysis techniques; personal direct contacts with the local management and technological tests of applicable programs.

The success of the results obtained, found support in the public organization, more specifically in the opening of the legislation of the local political organs, of the area of Guimarães - There was a readiness for dialogue towards the applicability of the material of the data collection and a good response.

Application and Managerial Implication

The contribution of the research is an initial reflection about mentioning a structural analysis in information centres to delimitate the ideal architecture so they can be transformed into knowledge and value generating centres through their management (Mohssin and Al-Ahmad, 2005).

Being so, it crosses the experience of the development of knowledge centres.

We look forward to identifying features and characteristics common to the organizational theories of the administrative management, with important implications to the organization of knowledge. Hasitschka et al (2005) considers in order to develop a reference or model which would constitute a relevant instrument for those who share the management of libraries.

The Raul Brandão Library has been using their produced knowledge in the optimization of their actions. The importance given to the new administrative management practise is the power to transform information into knowledge, on the way to the production of institutional value. This centre of knowledge understand their value in the market and use mechanisms: the action of their professionals in the attention given to the market in which they are inserted; cooperation of other libraries which contribute to answering local needs (Figure 3); advising for the acquisition of information resources. Forming an image of identity in the knowledge produced.

Figure 3. Planning of the Raul Brandão Library's

However, it is necessary to know the different skill combinations. When mobilizing human skills, people will feel better in relation to their activities performed in the organization and, therefore, will produce with greater competence and with a higher level of commitment with the results that could be reached.

They need the continuity of the analysis of the formation of the administrative theory, since the impact on social progress that the organizational environment has obtained, is better understood through the programs of bringing value to knowledge. For its applicability, there is the need of analysis of the human resources structures, available in libraries.

Conclusion

With the evolution of the organizations theory, new management practices have transformed the institutions. The improvement of these management practices in libraries is constant, and interacts in an international axis, since each unity increasingly communicate with other.

As a reflection of organizational administrative practices, libraries have developed and assumed the organizational theories to make a framework easily for effective management practices. Comprehending the importance of the studies on libraries requires both a new approach of existing public policies for this field, and the recognition of the Information Science knowledge on area.

In this way, it was planned to develop a model constituting a tool for managers in these institutions through the experience of laboratory works carried out with researchers from the European community. Portugal libraries are institutions that operate in the social process of their communities and the archives establish their structural spaces in parallel with the political, economic and social environments of the communities, reflecting the society and its organic sphere of informational performance.

The information plays a cyclical and essential role for the arrangement of the effective and necessary record to the organizational process and the institutional growth. Studies of this nature will help to valuate an economy based on knowledge management, and its production emphasizes the human capital that permeates the condition of the information in support for the development of communities and the growth of the organizational environment in political, social and cultural spheres.

Acknowledgment

This research was supported by CAPES (Coordination for the Improvement of Higher Education Personnel) and your publication was supported by FUNDUNESP (Foundation for the Development of the Unesp)

References

Argyris, C. & Schon, D. A. (1978). 'Organizational Learning: A Theory of Action Perspective,' *Addison-Wesley, Reading, MA.*

Bateman, T. S. & Snell, S. A. (1998). Management. England: Academic Internet Publishers.

Costa, M. D., Krucken, L. & De Abreu, A. F. (2000). "Gestão da Informação ou Gestão do Conhecimento," *Revista ACB: Biblioteconomia em Santa Catarina.* 5:5.

Chiavenato, I. (2000). Management 3. ed. São Paulo: Makron.

Drucker, P. F. (1994). Post-Capitalist Society, *Harper Business*: New York.

General Report on the Activities of the European Union. Artigos 212° do Tratado CE e 125.° do Tratado CEEA. European Parlament. February, 2006.

Hasitschka, W., Tschmuck, P. & Zembylas, T. (2005, Summer). "Cultural Institutions Studies: Investigating the Transformation of Cultural Goods," *The journal of Arts Management, Law, and Society.* 35 (2). 147-158.

Kaplan, R. S. & Norton, D. P. (2000). The Strategy-focused Organization: How Balanced Scorecard Companies Thrive in the New Business Environment, London: *Harvard Business School.*

Mohssin, I. & Al-Ahmad, N. (2005). "The Role of Information Technology in Building up Knowledge Economy Producting and Using Information in Libraries and Information Centers," *Journal of Social Sciences*, 1(4) 203-210.

Nonaka, I. & Konno, N. (1998). "The Concept of 'Ba': Building a Foundation for Knowledge Creation," *California Management Review* 40(3), 40-54.

Yin, R. K. (1989). Case Study Research: Design and Methods, *Sage Publications Inc.,* USA.

7

Intranet Supported Knowledge Sharing Behavior

Mohamad Noorman Masrek, Hasnah Abdul Rahim,
Rusnah Johare and Yanti Rahayu Rambli

Universiti Teknologi MARA, Shah Alam, Malaysia

Abstract

While most corporate organizations in Malaysia have implemented intranet or portal, questions regarding users' utilization behavior for the purpose of knowledge sharing still remain unanswered. Against this concern, this study seeks to investigate demographic profiles associated with knowledge sharing behavior in an intranet computing environment in selected Malaysian companies. Using the survey research method, 700 questionnaires were distributed using the simple random sampling technique yielding to 359 usable responses. The findings suggest that there is a significant difference in terms of knowledge sharing behavior across demographic profiles. In addition, it was also found that both length of service and Internet experience is a significant predictor of knowledge sharing behavior in an intranet computing environment.

Keywords: knowledge sharing, intranet, Malaysian companies

Introduction

Since its first inception a decade ago, the intranet has achieved major advancement and sophistication. At present, intranet technologies have significantly mature and they exist in all sizes, shapes, and forms. In fact, more sophisticated terms like intranet portal, enterprise portal, enterprise information portal or EIP (Shilakes and Tylman, 1998) have been coined to reflect the advancement and complexity of the technology. Hinrichs (1997) defined intranet as an internal IS based on internet technology, web services, TCP/IP and HTTP communication protocols, and HTML publishing that permits organization to define itself as a whole entity, a group, a family, where everyone knows their roles and everyone is working on the improvement and health of the organization.

Intranet, in its full functionalities can be used as a publishing application, discussion application and interactive application. Within the scope of discussion application, users can utilize the intranet technology for knowledge sharingpurposes. Today, while most corporate organizations in Malaysia are intensely implementing intranet or portal, questions regarding users' utilization behavior for the purpose of knowledge sharing still remain unanswered. Not much is really known about the extent Malaysian users utilize intranet technology for knowledge sharing purposes. Against this background, this study seeks answer the following research questions i.e. What are the demographic profiles associated with knowledge sharing behavior in an intranet computing environment? In addition, it also attempt to find answers to the following: (i) Is there any significant difference of knowledge sharing behavior between male and female? (ii) Is there any significant difference of knowledge sharing behavior among different age group of users? (iii) Is there any significant difference of knowledge sharing behavior between managers and non-managers? (iv)Does length or service significantly and positively relate to knowledge sharing behavior? and (iv) Does internet experience significantly and positively relate to knowledge sharing behavior?

Literature Review

Intranet Background

While the Internet started out from the ARPANET project in the late 1960s, intranets are the result of the growing number of companies beginning to run TCP/IP on their intra-organizational networks in the mid-1990s (Slevin, 2000). Karlsbjerg and Damsgaard (2001) described intranet as "a shared information space that supports the sharing of information among members of an organization. The space comprised a number of technical standards and platforms interconnected in a network within well-defined boundaries of a group of people or computers. All communication goes through the web-browser using TCP/IP and HTTP protocols. Thus, any application can be part of the intranet as long as the browser is primary client interface". Intranets are also sometimes referred to as 'glueware' or 'middleware' since they are utilized to interconnect heterogeneous systems through the browser and associated protocols and applications (Lyntinen et al., 1998).

Looking from the IS perspectives, intranet technologies offer formidable benefits compared to traditional technologies which tend to only support well-defined tasks (Damgaards and Scheepers, 1999). These advantages include rapid scalable development across a range of platforms, access to legacy systems and data warehousing capabilities, and development on existing networks with lower

implementation cost compared to traditional client server solutions (Golden and Hughes, 2001). In addition, the time taken for intranet implementation which includes design, development and implementation and end-user training is relatively much quicker than traditional solutions. Hence, the intranets are providing organizations with far more flexibility than traditional IS (Golden and Hughes, 2001).

Intranet Usage for Knowledge Sharing

Recognizing the importance of knowledge sharing, many organizations have deployed or exploited the intranet as part of their knowledge management initiative programs. The literature indicates that there exist diverse studies that specifically address the role of intranet in facilitating knowledge sharing (Newell et al., 1999; Ruppel and Harrington, 2001; Stoddart, 2001; Holden, 2003; Lichtenstein et al., 2004; Hall, 2004; Panteli et al., 2005; Stenmark, 2005c; Stenmark, 2005e). Other studies such as Scott (1998); Stenmark (1999a); Damsgaard and Scheepers (2001); Stenmark (2002); Sarkar and Bandyopadhyay (2002); Dingsoyr and Conradi (2003); and Skok and Kamanovitch (2005) not only addressed primarily the role of the intranet in supporting knowledge management initiatives but also stressed equal emphasis on knowledge sharing.

In order to best describe how the intranet can facilitate knowledge sharing, Stenmark (2002) and Lichtenstein et al., (2004) developed a model that describes intranet utilization for supporting knowledge management. Stenmark's model, suggests that the intranet as a knowledge sharing environment is seen from three perspectives: information, awareness and communication. The information perspective relates that the intranet gives the organizational members access to both structured and unstructured information in the form of databases and documents. Access to rich and diverse information is imperative for knowledge creation. The awareness perspective suggests that the intranet is used to keep users well-informed and constantly connected to information and people in the organization. Such a networking practice promotes community building and increases the likelihood for successful communication and collaboration. The communication perspective enables organizational members to collectively interpret the available information by supporting various forms of channels for conversation and negotiations. When users engaged in collaborative work with peers that share their objectives and understand their vocabulary, the common context for knowledge sharing would then exist.

Lichtenstein et al., (2004) conceptualization of knowledge sharing mediated by the intranet exhibits a sharer who chooses to provide knowledge to be published, and provides that knowledge which is then published on the intranet. A potential receiver will search and find the required knowledge, retrieve it then relate it to his/her existing knowledge. The knowledge is then assimilated before it can be applied as required. The fact that the knowledge has been retrieved by the receiver, as well as response to that knowledge, is fed back to the sharer, whose future knowledge-sharing behavior may change accordingly.

Individual Characteristics and Knowledge Sharing Behavior

Every individual is subject to his own personal traits and to the environment or surrounding that he belongs to or is attached with. Theory of Diffusion of Innovation (Rogers, 1995) posited that besides individual beliefs of the innovation characteristics (i.e. the object or technology being studied) other factors such as individual characteristics, organizational characteristics and external characteristics are also influential in molding one's behavior associated with individual adoption behavior. Models such as Technology Acceptance Model (Davis et al., 1989) and Unified Theory of Acceptance and Use of Technology or

UTAUT (Venkatesh et al., 2003) have been consistently showed by researchers that individual characteristics, organizational characteristics and technology characteristics are predictors or antecedents of technology adoption (see Jeyaraj et al., 2006). A large number of studies on the intranet have attempted to investigate the effects of the individual characteristics, organizational characteristics and technology characteristics on intranet adoption. However, these studies either done at the firm-level perspective (Al-Gharbi and Atturki, 2001; Eder and Igbaria, 2001) or user-level perspective (Horton et al., 2001; Weitzel and Hallahan, 2003; Chang, 2004) were meant to determine use or non-use and not for knowledge sharing behavior. Furthermore, in the context of Malaysia, none has ever attempted to investigate knowledge sharing behavior in an intranet computing environment.

Research Methodology

The conduct of the study involved survey research method. Several companies with high intranet maturity (i.e. the intranet are being integrated with organizational information systems) were contacted to participate in the survey. However, only four companies were willing to participate in the study. After a lengthy discussion with the contact person of these companies, it was decided that the respondents of the study should be the executives in the headquarters only. The rationale being that, compared with the support staffs, the executives are the heavy users of the intranet. Accordingly, 700 questionnaires were administered to these participating companies using stratified random sampling. After one-month duration, 423 were returned but 359 were found usable. An instrument comprising of six-item measures adapted from De Vries et al. (2006) was used to gauge knowledge sharing behavior. Data were analyzed using SPSS version 14.0. Non-response biases were analyzed by comparing early responders and late responders using independent sample t-test. The results revealed that the responses were free from non-response biases. Factor analysis was then executed on the items measuring knowledge sharing and the findings showed that all items were cleanly loaded into one single factor. The reliability analysis performed also showed that items measuring knowledge sharing recorded Cronbach alpha value of 0.907 suggesting that the instrument used in the study was highly reliable.

Findings

Demographic Profile

Table 1 presents the demographic profiles of the research samples. Between male and female, the former seemed to outnumbered the later with 54.9% as opposed to 45.1%. Age group between 31 and 35 was most dominant and contributed to 29.5% of the sample. In terms of qualifications, 284 respondents indicated to have gotten first

degree while 23 indicated to have obtained Masters. 306 respondents indicated as holding executives posts while 53 were holding middle management post. The average length of service was 7.62 while intranet experience recorded a mean of 6.92.

Knowledge Sharing Behavior between Gender

Table 2 depicts the descriptive profile of knowledge sharing behavior between male and female. The results showed that the mean score for both gender is around 5, suggesting that there is not much difference for both male and female in terms of knowledge sharing behavior. To further ascertain this finding, an independent sample t test was performed and the results evidently showed that the p value is 0.616 which is greater than 0.05, hence indicates that there is no significant difference on knowledge sharing between both gender.

Table 1. Demographic Profiles of Respondents

	Gender						
	Male				Female		
Frequency	197				162		
Percent (%)	54.9				45.1		
	Age						
	20 – 25	26 – 30	31 – 35	36 – 40	41 – 45	46 – 50	> 50
Frequency	33	91	106	85	33	6	5
Percent (%)	9.2	25.3	29.5	23.7	9.2	1.7	1.4
	Qualification						
	Diploma		Degree		Master	Others	
Frequency	28		284		23	24	
Percent (%)	7.8		79.1		6.4	6.7	
	Job Level						
	Executive				Mid. Mgt.		
Frequency	306				53		
Percent (%)	85.2				14.8		

Table 2. Descriptive Profile of Knowledge Sharing Behavior between Male and Female

	Gender	N	Mean	Std. Deviation	Std. Error Mean
Knowledge_Sharing	Male	197	5.0000	1.08849	0.07755
	Female	162	5.0586	1.11803	0.08784

Different Age Group of Users

To investigate whether there are significant differences across different age groups in terms of knowledge sharing behavior, ANOVA test was performed. The results showed that the p value is greater than 0.05, hence, suggesting that there is significant difference on knowledge sharing among different age group suggesting. Further analysis was performed using Scheffe test and the result is shown in Table 3.

Table 3. Results of Scheefe Test across Age Groups

	N	Subset for alpha = .05	
		1	2
Between 20 and 25	33	4.1313	
Between 26 and 30	90	4.7037	4.7037
Between 31 and 35	107	5.1121	5.1121
Between 36 and 40	85	5.3471	5.3471
Between 41 and 45	33	5.4646	5.4646
Between 46 and 50	6		5.6667
Between 51 and 55	5		5.8000
Sig.		.052	.204

Means for groups in homogeneous subsets are displayed.
a Uses Harmonic Mean Sample Size = 15.234.
b The group sizes are unequal. The harmonic mean of the group sizes is used. Type I error levels are not guaranteed.

Knowledge Sharing Behavior between Managers and Non-managers

Table 4 depicts the descriptive profile of knowledge sharing behavior between executive and middle managers. The mean value of knowledge sharing for middle managers seems to be higher as compared to the executives. However, to further ascertain whether this difference is significant, an independent sample t test was performed. Evidently, the results showed that the p value is smaller than 0.05, thus, implying that the difference is significant.

Table 4. Descriptive Profile of Knowledge Sharing between Executive and Middle Managers

	Job Level	N	Mean	Std. Deviation	Std. Error Mean
Knowledge_Sharing	Executive	306	4.9297	1.04326	0.05964
	Middle Management	53	5.5849	1.25899	0.17294

Length of Service and Knowledge Sharing Behavior

Table 5 and 6 present the results of linear regression between length of service and knowledge sharing behavior. It was noted that the value of Pearson's r = 0.318 while R^2 = 0.101, with $F_{(1,357)} = 40.213$ and $p < 0.001$. These figures show that low correlation but weak relationship subsist and that length of service single-handedly explained 10.1% of the variation of knowledge sharing behavior.

Table 5. Summary of Regression Model between Length of Service and Knowledge Sharing Behavior

Model	R	R Square	Adjusted R Square	Std. Error of the Estimate
1	.318(a)	.101	.099	1.04501

a Predictors: (Constant), LgthService

Table 6. Coefficient for Regression Model between Length of Service and Knowledge Sharing Behavior

Model		Unstandardized Coefficients		Standardized Coefficients		
		B	Std. Error	Beta	t	Sig.
1	(Constant)	4.486	.102		44.196	.000
	LgthService	.071	.011	.318	6.341	.000

a Dependent Variable: Knowledge_Sharing

Internet Experience and Knowledge Sharing Behavior

Table 7 and 8 depict the results of linear regression between length of service and knowledge sharing behavior. Tt was noted that the value of Pearson's r = 0.445 while R^2 = 0.198, with $F_{(1,357)}$ = 87.995 and $p < 0.001$. These figures show that low correlation but moderate relationship exist and that intranet experience singularly explained 19.8% of the variation of knowledge sharing behavior.

Table 7. Summary of Regression Model between Intranet Experience and Knowledge Sharing Behavior

Model	R	R Square	Adjusted R Square	Std. Error of the Estimate
1	.445(a)	.198	.195	.98731

a Predictors: (Constant), IntranetExp

Table 8. Coefficient for Regression Model between Intranet Experience and Knowledge Sharing Behavior

Model		Unstandardized Coefficients		Standardized Coefficients		
		B	Std. Error	Beta	T	Sig.
1	(Constant)	3.910	.130		30.105	.000
	IntranetExp	.161	.017	.445	9.381	.000

a Dependent Variable: Knowledge_Sharing

Discussion

Utilizing intranet for knowledge-sharing purposes has been vastly discussed both in IS and KM literature. Typically, knowledge-sharing activities entail two main activities namely; knowledge donating and knowledge collecting. With the availability of the intranet, these two activities have become more conveniently practiced. Users at any time of the day and at their own will can easily record or donate whatever knowledge that could be relevant and beneficial to others. On the other hand, users can also freely retrieve or collect knowledge pertinent to their needs from the intranet, thereafter assimilating them for usage or application. As this study had evidently showed that intranet was utilized for knowledge sharing purposes, findings by earlier studies are therefore, affirmed (Stenmark, 2002; Lichtenstein et al., 2004).

Numerous studies have consistently shown that individual characteristics such as demographic profile and other traits such as job level, length of service and internet experience are predictors of IT or IS usage behavior. While previous studies have found gender differences in terms of IT usage behavior (Gefen and Straub, 1997; Gardner, 2004) this study has, however, discovered contradicting result. The possible explanation could be that, unlike previous studies, IT usage behavior in this study was measured in terms of their knowledge sharing via the intranet. Furthermore, the job nature of these respondents, which are identical irrespective of their gender, could be also another reason.

This study has categorized the age of the respondents into 7 groups. As noted in the previous section, other than those aged between 20 and 25, other age groups displayed an almost identical pattern in terms of their knowledge sharing behavior. Compared to other age groups, respondents aged between 20 and 25 have the least number of experience working in the organizations. Being more junior, surely they

have fewer companions or cohort in the workplace and that would certainly limit their knowledge sharing activities.

With regards to knowledge sharing behavior in terms of job level, this study has found there is a significant difference between managers and non-manager (executives). However, this finding has to be interpreted with cautious considering uneven number of respondents between the two job levels. Nevertheless, it was clear that the overall mean for the middle manager is greater than the executives which could be attributed by the fact that the job nature of the manager requires them to repetitively share work knowledge especially with their subordinates. Other possible explanation would be that managers are the centers or nucleus for subordinates in their job reporting. In the process, these subordinates would be likely to share their knowledge and experience with their superior. Previous studies indicate that length of service has mixed results as a determinant of IS usage behavior. For instance, Burkhardt (1994) discovered negative correlation between length of service and computer usage while Liao and Landry (2000) found that staff with longer length of service tend to perceive the newly implemented IS as being very useful, which in turn became the strong predictor to the IS acceptance. Apparently, in terms of knowledge sharing behavior, the findings of this study is consistent with that of Liao and Landry (2000). Employees with longer length of service usually have better understanding of organizational processes and operations, as well as better involvement and contribution during the implementation of the intranet. Hence, these employees would perhaps perceive the intranet as being very useful, which in turn heightens usage level which includes the purpose of knowledge sharing.

In investigating individual IS utilizaton behavior, many researchers had also studied the effect of computer or internet experience. Evidently, many of these studies had found a strong support on the assertion that computer or internet experience is a predictor of IS usage behavior (Igbaria and Iivari, 1995; Hubona and Geitz, 1997; Alshare et al., 2004). In the context of intranet usage study, Chang (2004) also discovered similar finding, which is also confirmed by the findings of this study. Intranet is almost similar to internet and the only difference is on the breadth and scope of its user coverage. Therefore, users who are already familiar with the internet and web-applications will find the intranet is just as convenient to the internet which in turn promote their usage level not only for information searching but also for knowledge sharing.

Conclusion

The conduct of this study has been to investigate the demographic profiles of knowledge sharing behavior in an intranet computing environment among executive staffs in selected Malaysian. The study has provided empirical evidence on the importance of intranet as an important knowledge sharing tool in the workplace. Despite the success of accomplishing the research objectives, this study is also subject to a number of limitations which is mainly associated with adopted research method. Firstly, the chosen respondents were those holding executive level positions or higher, and users of the lower level positions were omitted. Therefore, future research should consider adopting every intranet user irrespective of their job level as respondents. As such, differences and comparisons can be made between job-level and status of intranet utilization. Secondly, the chosen companies were of those of GLC only, hence other public and fully private companies were excluded. Among government organizations, little is really known about the intranet utilization behavior among civil servants. Apparently, it is worth venturing into research that explores the status of intranet utilization in government agencies especially when these organizations are steadily geared towards paperless office or e-

government. Equally appealing would be investigating the status of intranet utilization in fully private companies such as those from banking or manufacturing industries. Undeniably, the situation and atmosphere in these companies are totally different as those from the GLC or government as they are more tailored towards cost-saving and profit making. Apparently, this situation warrants research undertakings. Thirdly, the perceptual measures employed in the survey instrument are subject to individual interpretation and understanding. Hence, instead of using self-reported measures for measuring intranet usage for knowledge sharing, a more accurate approach would be installing software-tracking systems onto the intranet that would both monitor and record usage. However, such approached would be quite difficult unless permission and access are granted by the organization's intranet. Also, objective measures as a substitute to perceptual measures would also provide more accurate measurement.

References

Al-Gharbi, K. & Atturki, S. M. (2001). "Factors Influencing the Infusion of Intranet in Knowledge Management in Developing Countries Organizations," Proceedings of the 10th International Conference on Management of Technology, Lausanne, Switzerland.

Alshare, K., Grandon, E. & Miller, D. (2004). "Antecedents of Computer Technology Usage: Considerations of the Technology Acceptance Model in the Academic Environment," *Journal of Computing Sciences in Colleges*, 19(4), 164–180.

Burkhardt, M. E. (1994). "Social Interaction Effects Following a Technological Change: A Longitudinal Investigation," *The Academy of Management Journal*, 37, 869–896.

Butler, P., Cales, R.& Petersen, J. (1997). Using Microsoft Commercial Internet System, Indianapolis, *Que Publishing*.

Casselberry, R. et al. (1996). Running a Perfect Intranet, Indianapolis, *Que Publishing.*

Chang, P. V. (2004). "The Validity of an Extended Technology Acceptance Model (TAM) for Predicting Intranet/Portal Usage," *Unpublished masters dissertation*, University of North Carolina at Chapel Hill, USA.

Damsgaard, J. & Scheepers, R. (2001). "Using Intranet Technology to Foster Organizational Knowledge Creation," *Global Co-Operation in the New Millennium*, Proceedings of the 9th European Conference on Information Systems (ECIS2001), Bled, Slovenia

Davis F. D., Bagozzi, R. P. & Warshaw, P.R. (1989). "User Acceptance of Computer Technology: A Comparison of Two Theoretical Models," *Management Science*, 35 (8), 982–1003.

De Vries, R. E., Den Hooff, B. V. & De Ridder, J. A. (2006). "Explaining Knowledge Sharing: The Role of Team Communication Styles, Job Satisfactions and Communication Beliefs," *Communications Research*, 33(2), 115–135.

Dingsoyr, T. & Royrvik, E. (2003). "An Empirical Study of an Informal Knowledge Repository in a Medium-Sized Software Consulting Company," Proceedings of the 25th International Conference on Software Engineering (ICSE2003), Portland, Oregon, USA.

Eder, L. B. & Igbaria, M. (2001). "Determinants of Intranet Diffusion and Infusion," *Omega*, 29 (3), 233–242

Gardner, C. & Amoroso, D. L. (2004). "Development of an Instrument to Measure the Acceptance of Internet Technology by Consumers," Proceedings of the 37th Hawaii International Conference on System Sciences – 2004.

Gefen, D. & Straub, D. W. (1997). "Gender Differences in the Perception and Use of E-Mail: An Extension to the Technology Acceptance Model," *MIS Quarterly*, 389-400.

Golden, W. & Hughes, M. (2001). "Business Process Re-Engineering Using Intranets: A New Beginning," Proceedings of the 14th Bled Electronic Commerce Conference, Bled, Slovenia.

Hall, H. (2004). "The Intranet as Actor: The Role of the Intranet in Knowledge Sharing," Proceedings of the International Workshop on Understanding Sociotechnical Action (USTA2004), Edinburgh, Scotland, 109 –111.

Hinrichs, R. J. (1997). "Intranets: What's the Bottom Line?," *Mountain View, CA, Sun Microsystems Press.*

Holden, T. (2003). "Understanding the Dimensions of Knowledge Sharing: Designing an Intranet to Improve Operational Performance in a Multinational Corporation," *International Journal of Electronic Business.* 1(2), 118–139.

Horton, R. P., Buck, T., Waterson, P. E. & Clegg, C. W. (2001). "Explaining Intranet Use with Technology Acceptance Model," *Journal of Information Technology*, 16, 237–249.

Hubona, G. S. & Geitz, S. (1997). "External Variables, Beliefs, Attitudes and Information Technology Behavior," Proceedings of the 30th Annual Hawaii International Conference on System Sciences (HICSS97), Hawaii, USA.

Igbaria, M. & Iivari, J. (1995). "The Effects of Self-Efficacy on Computer Usage," *Omega*, 23(6), 587–605.

Jeyaraj, A., Rottman, J. W., & Lacity, M. C. (2006). "A Review of Predictors, Linkages and Biases in IT Innovation Adoption Research," *Journal of Information Technology*, 21, 1–23.

Karlsbjerg, J. & Damsgaard, J. (2001). "Make or Buy – A Taxonomy of Intranet Implementations Strategies," Proceedings of the 9th European Conference on Information Systems (ECIS2001), Bled, Slovenia.

Liao, Z. & Landry, R. (2000). "An Empirical Study on Organizational Acceptance of New Information Systems in a Commercial Bank Environment," Proceedings of the 33rd Hawaii International Conference on System Science (HICSS2000), Hawaii, USA.

Lichtenstein, S., Hunter, A. & Mustard, J. (2004). "Utilization of Intranets for Knowledge Sharing: A Socio-Technical Study," Proceedings of the 15th Australasian Conference on Information Systems (ACIS2004), Tasmania, Australia.

Lyntinen, K., Rose, G. & Welke, R. (1998). "The Brave New World of Development in the Internetwork Computing Architecture (Internca) or How Distributed Computing Platforms Will Change Systems Development," *Information Systems Journal*, 8, 241–253.

Newell, S., Scarbrough, H., Swan, J. & Hislop, D. (1999). "Intranets and Knowledge Management:
Complex Processes and Ironic Outcomes," Proceedings of the 32nd Hawaii International Conference on System Sciences (HICSS1999), Hawaii, USA.

Panteli, N., Tsiourva, I. & Modelley, S. (2005). "Intra-organizational Connectivity and Interactivity with Intranets: The Case of Pharmaceutical Company," *Working paper*, University of Bath, UK.

Rogers, E. M. (1983). Diffusion of Innovations. New York: *Free Press*.

Ruppel, C. P. & Harrington, S. J. (2001). "Sharing Knowledge through Intranets: An Analysis of the Organizational Culture Leading to Intranet Adoption and Use," *IEEE Transactions on Professional Communications*, 44(1), 37.

Sarkar, R. J. & Bandyopadhyay, S. (2002). "Developing an Intranet-Based Knowledge Management Framework in a Consulting Firm: A Conceptual Model and its Implementation," *Workshop on Knowledge*

Management and Organizational Memories (ECAI02), Lyon, France.

Scheepers, R. (1999). "Intranet implementation: Influences, Challenges and Role Players," *Unpublished doctoral dissertation*. Aalborg University, Denmark.

Scott, J. E. (1998). "Organizational Knowledge and Intranet," *Decision Support Systems*, 23, 3–17.

Shilakes, C., & Tylman, J. (1998). 'Enterprise Information Portals,' New York, *Merrill Lynch, Inc.*

Skok, W. & Kamanovitch, C. (2005). "Evaluating the Role And Effectiveness Of An Intranet In Facilitating Knowledge Management: A Case Study At Surrey County Council," *Information & Management*, 42, 731–744.

Slevin, J. (2000). The Internet and Society, Cambridge, *Polity Press*.

Stanek, W. R. (1997). Web Publishing: Professional Reference Edition Unleashed, Indianapolis, Sams.net Publishing.

Stenmark, D. (1999a) "Using Intranet Agents to Capture Tacit Knowledge," Proceedings of World Conference on the WWW and Internet, (WebNet99), Honolulu, Hawaii, USA, 1000 – 1005.

Stenmark, D. (2002). "Information vs. Knowledge: The Role of Intranets in Knowledge Management," Proceedings of the 35th Hawaii International Conference on System Sciences (HICSS2002), Hawaii, USA.

Stenmark, D. (2005c). "Knowledge Sharing on a Corporate Intranet: Effects of Re-Instating Web Authoring Capabilities," Proceedings of the 13th European Conference on Information Systems (ECIS 2005), Regensburg, Germany.

Stenmark, D. (2005e). "Knowledge Sharing through Increased User Participation on a

Corporate Intranet," Proceedings of the 6th European Conference on Organizational Knowledge, Learning, and Capabilities (OKLC2005), Bentley College, Massachusetts, USA.

Stoddart, L. (2001). "Managing Intranets to Encourage Knowledge Sharing: Opportunities and Constraints", *Online Information Review*, 25(1).

Venkatesh, V., Morris, M. G., Davis, G. B. & Davis, F. D. (2003). "User Acceptance of Information Technology: Toward a Unified View," *MIS Quarterly*, 27(3), 425–478.

Weitzel, D. & Hallahan, K. (2003). "Organizational Adoption of an Intranet-Based Performance Reporting System: A Test Of Rogers' Model of Innovation," In the New Technologies in Organizational Contexts Communications and Technology Division, International Communication Association, San Diego, 26th May 2003 [On-line] [Retrieved 7 June, 2005] http://lamar.colostate.edu/~pr/adoptionepower.doc

Zachman, J. A. (1987). "A Framework for Systems Architecture," *IBM Systems Journal*, 26 (3), 276 – 292.

Zimmerman, S. & Evans, T. (1996). Building an Intranet with Windows NT 4. Indianapolis, Sam.net Publishing.

8

A Swap of Perspectives: Data Migration and Knowledge Management as Mutually Interrelated Disciplines

Peter Kažimír[1], Vladimír Bureš[1,2] and Tereza Otčenášková[2]

[1]College of Management, Bratislava, Slovakia

[2]University of Hradec Králové, Hradec Králové, Czech Republic

Abstract

Development of business environment is strongly influenced by quick penetration of information technologies (IT) into all organizational processes. This process generates several problems, which single organizations need to cope with. A specific problem is closely related to vast amount of data which is produced every day, processed to valuable information and consequently applied during decision-making processes. There are several reasons, such as progress in business informatics and the IT industry in general, that lead to necessity to change the used IT system completely. This change is accompanied by the requirement of data migration. This process is performed only occasionally, therefore the lack of experience can cause troubles such as lost data or its low usability in the new system. On the other hand, knowledge management offers tools and methods how to preserve experience and lessons learned and hence avoid several pitfalls of the data migration process. The paper investigates mutual interrelationship of two distinct areas, Data Migration processes and Knowledge Management programs. Basic facts together with various definitions and characteristics of both research topics are presented. The paper reveals that these are closely connected in terms of how data, information and knowledge should be created, enhanced and managed. Therefore, similarities and mutual features of both disciplines are outlined.

Keywords: Interrelationship, Data Migration, Knowledge Management, Perspectives.

Introduction

The pursuit of higher competitiveness makes companies continuously try to apply various types of business concepts and methodologies, or information technology tools, techniques, and methods. One of the business concepts widely implemented in companies is Knowledge Management. Its scope or form is determined by the objectives of organization and the personal objectives and needs of people inside it.

Generally, desired implications of Knowledge Management implementation are cost reduction, customer or employee satisfaction and quality (Spek and Kingma, 1999).

Knowledge Management principles are usually applied to business activities, which are knowledge intensive or demanding (Bureš and Čech, 2007). In the business informatics domain there are projects which companies have to execute and are unavoidable due to penetration rate of information technologies into the contemporary business environment. Information technology (IT) migration in general or Data Migration in particular can serve as an example of such knowledge intensive projects (Kantawala, 2008). On the other hand, Data Migration projects rely on quite significant body of knowledge, which is usually not managed due to low frequency of Data Migration projects occurrence. Therefore, these projects become often more complicated and expensive than previously anticipated (Haller, 2009). Hence, interconnection of both disciplines can be very fruitful and beneficial.

Search for described experience with both Knowledge Management and Data Migration in scientific databases (e.g. Springer, or Elsevier) returns only journal papers or book chapters focused on each discipline separately. However, few studies implicitly tie Knowledge Management and particular aspects of Data Migration. For instance, general frameworks of the Data Migration projects can be considered as outputs based on certain volume of knowledge (Jing et al, 1998; Aboulsamh and Davies, 2011). Nevertheless, this connection is not explicitly stated. Therefore, there is a necessity for mutual interrelation of both domains and explicit identification of mutual features. This endeavor can be contributive to development of both theory and practice of Knowledge Management and Data Migration.

This paper is divided into three sections. First section provides the basic overview of Data Migration issues, including definition, or methodology of its execution. The second section describes Knowledge Management, its characteristics, features and the basics of its implementation. In the third section the relationship between Knowledge Management and Data Migration is issued. It contains a discussion on relationship between both projects, how they can help each other during implementation and comparison of project's methodologies and best practices from both areas of expertise.

Data Migration

There are various IT migrations projects realized by companies – application migration (CISCO, 2009), business process migration (Aversano et al., 2003), or data center migration (EMC, 2011). Since in the IT world data represents the basic building blocks, all aforementioned types of IT migration have to appropriately cope with Data Migration issues, i.e. IT migration projects have to ensure that correct data from the old system will be preserved and transferred to a new system. Data Migration can be considered as the process of transferring data between storage types, formats, databases applications or computer systems (Bartkus, 2011). In order to have proper data in the new system, following high level activities need to be executed: planning, analysis and design, implementation and closeout (FSA, 2007). All these activities should have scheduled exact timeliness of their occurrence.

As we can see from the published case studies, Data Migration is unavoidable and it can get messy, time-consuming and difficult to conduct. Some surveys quote that 84 % of Data Migration projects running late, over budget or both (Howard, 2011). Data migration is a difficult and unattractive task with high potential for failure. There are five significant barriers to data migration success (Bell, 2011):

- Delaying the data migration effort until it adversely affects the system conversion effort.

- Failing to make informed data migration decisions due to lack of cost and time estimates.

- Failing to fully engage the business in the data migration project.

- Inability to access scarce internal subject matter experts.

- Using inexperienced staff with homegrown tools and unproven processes.

To prevent project failure there have been already surveys, research studies and methodologies created (FSA, 2007), (Howard, 2011), (QLOGIC, 2008), or (Manek, 2003) – see example in Table 1. In summary to ensure successful Data Migration companies need to focus on following nine critical success factors:

1. Perform Data Migration as an independent project, ranging from budgeting through to testing.

2. Establish and manage expectations throughout the process.

3. Understand current and future data and business requirements.

4. Identify individuals with expertise regarding legacy data.

5. Collect available documentation regarding legacy system(s).

6. Define Data Migration project roles and responsibilities clearly.

7. Perform a comprehensive overview of data content, quality, and structure.

8. Adopt a formal methodology that has been tried and tested.

Coordinate with business owners and stakeholders to determine importance of business data and data quality.

Table 1: A High Level Approach to Data Migrations

Stage 1 – Define migration approach	Stage 2 – Plan and conduct data cleansing	Stage 3 – Design migration system	Stage 4 – Construct and unit test migration systems	Stage 5 – Conduct system test	Stage 6 – Convert data into production
Confirm objectives and scope	Prioritise datasets for assessment	Identify data conversion tools	Develop / deploy the extract, transform and load software	Establish data migration testing mechanisms	Load reference data
Develop schedule	Define data quality criteria	Develop data reconciliation approach			Ensure all data cleansing complete
Identify stakeholders	Identify data quality tools				
Establish conversion roles and data quality committee	Assess source system data	Create migration system architecture	Unit test migration system	Conduct data migration trial	Conduct full production migration
Define activities, dependencies and deliverables	Review assessment with data quality decision committee				
Kick-off project		Create automated data cleansing specification	Identify and resolve software issues	Identify and resolve software issues	Resolve data quality and reconciliation issues
Develop data models and data mappings	Cleanse data				
Project management					
Planning		Communication		Risk Management	Quality Assurance

Knowledge Management

Although there are already a number of documented achievements, many companies are reluctant to undertake Knowledge Management initiatives because of the difficulty in establishing a sound business case. The difficulty in establishing a business case for Knowledge Management programs is complexity, mostly handled with a method of trial and error. It often stems from the fact that, since knowledge is intangible, it is difficult to clearly see direct link from a Knowledge Management process to a demonstrable business outcome (Yelden and Albers, 2004).

Knowledge Management involves a strategic commitment to improving the organization's effectiveness, as well as to improving its opportunity enhancement. There are three spheres of Knowledge Management (Pee and Kankanhalli, 2009):

- Technology – It provides a secure central space where employees, customers, partners and suppliers exchange information, share knowledge and guide each other and the organization to better decisions. This could be in the form of knowledge-portal on the corporate intranet or a centralized repository which allows the team members to use and share information.

- Knowledge Management processes – These include standard processes of knowledge contribution, content management, retrieval, membership on communities of practice, implementation projects based on knowledge reuse, methodology and standard formats to document best practices and case studies, etc.

- People – The biggest challenge in Knowledge Management is to ensure participation by all team members in knowledge sharing, collaboration and reuse to achieve business results. This is achieved by making small changes in the culture through combination of trainings, motivation/recognition and rewards etc.

The goal of Knowledge Management as a process is to improve the organization's ability to execute its core processes more efficiently. In order to achieve this goal Knowledge Management is formed as a set of proactive activities to support an organization in creating, assimilating, disseminating, and applying its knowledge. In addition, Knowledge Management is not one time job but it is a continuous process to understand the organization's knowledge needs, the location of the knowledge, and how to improve the knowledge (Hussain et al., 2004). As an interest in Knowledge Management and organizational knowledge grows IT researchers have been promoting a class of information systems, referred to as Knowledge Management Systems (Alavi and Leidner, 2001). In general there are four broad objectives of Knowledge Management Systems in practice (Davenport et al., 1998):

- Create knowledge repository.

- Improve knowledge assets.

- Enhance the knowledge environment.

- Manage knowledge as an asset.

All four of them try to bridge the gap between tacit and explicit knowledge. In practice, any technological solution that could assist in this process is highly appreciated (Stenmark, 1999). One of the keys of the Knowledge Management System is in the way of how to capture intellectual assets for the tangible benefits for the organization. As such, imperatives of Knowledge Management are to:

1. Transform knowledge to add value to the processes and operations of the business.

2. Leverage knowledge strategic to business to accelerate growth and innovation.

3. Use knowledge to provide a competitive advantage for the business.

One of the main goals of all three instances is to ensure that the right knowledge is delivered to the appropriate place or competent person at the right time to enable competent decision making. Applied knowledge has to be subjected to and pass tests of validation (Firestone, 1998). Moreover, the decision should be made based on knowledge which is supported by appropriate information. Appropriate information needs to be extracted, filtered, or formatted in a specific way.

Discussion

The previous two sections introduced Data Migration and Knowledge Management concepts and outlined their definitions, characteristics, typical features, and processes. They are used as a starting point for discussion in the following section, in which a connection between Data Migration and Knowledge Management is explained and areas where both can help each other to make execution easier and smoother are depicted.

Knowledge Management Features in Data Migration

Second section outlined nine critical success factors of successful Data Migration projects. Not surprisingly, they can be linked with features of Knowledge Management. The explanation is provided in this subsection.

Perform Data Migration as an Independent Project, Ranging from Budgeting through to Testing

Since Data Migration is a project which is unavoidable (QLOGIC, 2008) then lessons learned should be created during the project execution for future reference. Obviously, the lessons learned from previous projects should be always revisited at the beginning of every project (PRINCE2, 2009). Knowledge Management can help to create and use high-

quality lessons learned, which basically captures a knowledge that is applied to future action and derived from screening according to specific criteria (Patton, 2001):

- Evaluation findings — patterns across programs;

- Basic and applied research;

- Practice wisdom and experience of practitioners;

- Experiences reported by program participants/clients/intended beneficiaries;

- Expert opinion;

- Cross-disciplinary connections and patterns;

- Assessment of the importance of the lesson learned; and

- Strength of the connection to outcomes attainment.

The idea of high quality lessons learned is that the greater the number of supporting sources for a "lesson learned", the more rigorous the supporting evidence, and the greater the triangulation of supporting sources, the more confidence one can have in the significance and meaningfulness of a lesson learned.

Establish and Manage Expectations throughout the Process

Previous experience embodied in lessons learned can be applied as a foundation for expectations setting. They can be further shaped by knowledge of experienced employees who can contribute to the formulation of expectations. The significance of establishment of expectations is strengthened by the fact that this factor is important for both types of discussed projects. Therefore, generalization or analogy can be used during this process, and hence basic ideology can be shared. The advantage of the idea to have shared expectations between both implementations is in the centralized manipulation of data, information and knowledge. It will ensure that the same activity is not repeated for both projects separately.

Understand Current and Future Data and Business Requirements

Data is always created in a specific context and, if it is digital, then it is always created by a program of one kind or another. Some data is created for local use. A sales order system records sales orders and that data may be used elsewhere in the order-to-cash system of which it is a part (Bloor, 2001). Knowledge Management can help to analyze the data and acquire better understanding of what information data contain and identify individuals who benefits from getting this knowledge, and information. Consequently, the knowledge can enable better decision making and give better insight into future business requirements. Certainly, this approach has an impact on development of new IT infrastructure features, business applications and design of processes.

Identify Individuals with Expertise Regarding Legacy Data

It is an advantage if legacy data, which needs to migrate, are handled during migration not only by professionals who have technical expertise but also by professionals who knows what knowledge data can provide to business stakeholders. Therefore they can identify more accurately which data needs to migrate and in which form they should be stored or transformed.

Collect Available Documentation Regarding Legacy System(s)

Individuals, who work with legacy data and legacy systems, have a knowledge which needs to be available to ensure successful Data Migration. The collection of knowledge about legacy systems is not an easy task to perform. In this case Knowledge Management activities can help to ensure that appropriate information and knowledge are collected and used during Data Migration. The usefulness of knowledge should be guaranteed if four broad objectives of Knowledge Management systems have been met: create knowledge repository, improve knowledge assets, enhance the knowledge environment, and manage knowledge as an asset. Again, in this way Knowledge Management Systems can be used to help develop new IT infrastructure which will accommodate current and future business requirements.

Define Data Migration Project Roles and Responsibilities Clearly

Even the project will be executed by outside resources there should be developed at least some internal competency with respect to the project. Identification of an appropriate candidate with proper set of knowledge in the given area is crucial. Moreover, ethical issues are of high importance here (Semrádová and Kacetl, 2011). There should not be confusion about Data Migration project roles and their particular responsibilities. Based on the sphere "People" Knowledge Management can help to identify roles which ensure that the data, which are subject of the migration, are really the data which needs to migrate. It also can help to define responsibilities for the particular roles to ensure that knowledge is shared and prepared to be reused to achieve successful Data Migration.

Perform a Comprehensive Overview of Data Content, Quality, and Structure

It is desirable to start work early on understanding companies' legacy data assets and use the knowledge gained about the data assets to develop defendable time and cost estimates to attract management attention, resources and commitment (Bell, 2011). The knowledge about legacy data can help also to design a new IT infrastructure which will accommodate customers' and business needs for less cost. To achieve the best results it is advised to use data cleansing, data profiling and data integration tools. Broad analysis of data is not easy to accomplish as well as to create extensive overview of the outcome of analysis. Additionally the overview needs to be intelligible not only to IT staff but also to business staff and end users. The first objective of Knowledge Management System is to create knowledge repository which once created can help to understand data more easily and subsequently help to create an overview of their content, quality and structure.

Adopt a Formal Methodology that has been Tried and Tested

Methodology can be defined as a body of practices, procedures, and rules or study or theoretical analysis of working methods. Basically it provides already created knowledge about what tasks are needed to be done and the reasons why and how they need to be executed. The objective of Knowledge Management is not only to create knowledge but also to improve, enhance and manage knowledge. Consequently the benefit of Knowledge Management is in the way how methodology can be adjusted and how knowledge included in methodology can be used during Data Migration.

Coordinate with Business Owners and Stakeholders to Determine Importance of Business Data and Data Quality

The common feature of all Data Migration and Knowledge Management projects is that the business MUST be engaged throughout all stages of projects, from initial scoping to final completion (Bureš, 2006). One of the spheres of Knowledge Management is "People". Knowledge Management should ensure participation by all team members in knowledge sharing, collaboration and reuse to achieve business results. In the case of Data Migration it should ensure knowledge sharing, collaboration and reuse not only between project team members but also

between IT and Business (Jorfi et al., 2011) to achieve successful Data Migration.

Data Migration Aspects in Knowledge Management

Third section delineated essential features and characteristics of Knowledge Management. Four keys of successful Knowledge Management implementation were pointed out. All four of them can be linked with aspects of Data Migration. The explanation is provided below.

Create Knowledge Repository

The objective is to create repositories by storing knowledge and making it easily available to users. In order to store knowledge there is a need to create knowledge. It was pointed out in (Huber and O'Deil, 2000) that information and knowledge form a virtuous circle. Knowledge can be perceived as "information in use". Knowledge cannot exist without information. With good information, people can make better decisions and take intelligent action. As we already know information comes from data which in the case of Data Migration companies want to transfer to new environments such as databases, storage devices, computer systems, etc. While the data needs to be analyzed during Data Migration and likewise they need to be extracted, filtered, and formatted in a special way it would be beneficial to take advantage of this activity to support Knowledge Management programs to accelerate the speed of knowledge creation and transfer in the company (Coviello et al., 2001). One of the ways how to create knowledge could be via modeling from data (Abdullah et al., 2002). Consequently knowledge is stored in new environment where it should be easier accessed by end users as it was in old environment.

Improve Knowledge Assets

This objective expects that accesses are provided to knowledge and hence knowledge transfer is facilitated. In this case Data Migration can be seen as a strategy to transfer data, information and knowledge. To get the best out of it from the Knowledge Management perspective, a codification strategy with IT can be used to make the knowledge even more explicit. Consequently, the dissemination of knowledge throughout the organization can be performed quicker, by making it readily available in databases, decision support systems, expert systems, or recommendation systems (Čech and Bureš, 2007), ((Bloodgood et al, 2001).

Enhance the Knowledge Environment

The outcome of the objective should be an environment that encourages the creation, transfer and use of knowledge, regardless the specifics of particular environment (Mikulecký, 2003). Data Migration does not only transfer data from one data storage to another but also support and encourage implementation of new tools and systems which accommodate new business needs. Consequently, all systems and tools which are involved during Data Migration can help to design and support a Knowledge Management solution. Such typology consisting of tools as intranet systems, Electronic Document Management (EDM), groupware, workflow, artificial intelligence-based systems, Business Intelligence techniques (BI), knowledge mapping, competitive intelligence tools and knowledge portals were discussed already in terms of their potential contributions to the processes of creating, registering and sharing knowledge (Baroni de Carvalho and Ferreira, 2001).

Manage Knowledge as an Asset

Data Migration enables the rejuvenation of existing business systems and leverages application use, offering opportunities that current and future technologies provide (Syntel, 2006). For example the successful migration can ensure that business is fed by reliable and accurate reports which can be otherwise delivered in very time-consuming

process (Manion, 2001), (SkyParc, 2010). Likewise by adopting new service-oriented solutions, the interfaces of current applications can be updated to provide additional information, knowledge which can be managed as an asset on the balance sheet afterwards.

Conclusion

The paper has introduced specific information about two areas of expertise: Data Migration and Knowledge Management. It has provided definitions, processes, methodologies and nine success factors to ensure successful execution of Data Migration. It has covered also Knowledge Management topics, namely Knowledge Management goal, three Knowledge Management spheres, and objectives of Knowledge Management System. Afterwards the information about both topics has been used to find mutual beneficial characteristics which can help to make both implementation projects smoother and ensure their successful completion. The research showed that there are couples of features of Knowledge Management which can help Data Migration to be more successful and likewise there are several aspects of Data Migration which can help Knowledge Management to meet the strategic objectives. Even though results are promising, still detailed research needs to be done in this area. The deeper investigation of the provided ideas would be beneficial for both theoretical research in both areas and practical implementation of both types of projects. The research could also include a comparison of best practices from both areas.

Acknowledgement

This paper was created with the support of the Czech Science Foundation project SMEW, project number 403/10/1310.

References

Abdullah, M. S., Benest, I., Evans, A. & Kimble, C. (2002). Knowledge Modelling Techniques

For Developing Knowledge Management Systems, Proceedings of the 3rd European Conference on Knowledge Management, Dublin, Ireland, 15-25.

Aboulsamh, M. A. & Davies, J. (2011). "Specification and Verification of Model-Driven Data Migration," *Lecture Notes in Computer Science*, 6918, 214-225.

Alavi, M. & Leidner, D. E. (2001). "Knowledge Management and Knowledge Management Systems: Conceptual Foundations and Research Issues," *MIS Quarterly*, 25 (1), 107-136.

Aversano, L., Canfora, G. & De Lucia, A. (2003). Migrating Legacy system to the Web: A Business Process Reengineering Oriented Approach, Polo, M., Piattini, M. and Ruiz, F. (eds), Advances in Software Maintenance Management: Technologies and Solutions, *Idea Group Inc.*

Bartkus, G. (2011). 'Data Migration: Basic Overview,' Hie Electronics, McKinney, TX.

Bell, G. (2011). "Data Migration – Key Considerations," *The Data Administration Newsletter*, May 2011. [Online], [Retrieved April 4, 2012], http://www.tdan.com/view-articles/15145.

Bloodgood, J. M. & Salisbury, W. D. (2001). "Understanding the Influence of Organizational Change Strategies on Information Technology and Knowledge Management Strategies," Decision Support Systems, 31 (2001), 55-69.

Bloor, R. (2011). "What you don't Know about Data: What Should Data Know about Itself?," [Online], [Retrieved February 10, 2012], http://www.dataintegrationblog.com/robin-bloor/what-should-data-know-about-itself/.

Bureš, V. (2006). 'Knowledge Management and its Implementation,' Proceedings of the 2nd International Conference on Web Information Systems and Technologies, Setubal, Portugal, 115-118.

Bureš, V. & Čech, P. (2007). "Knowledge Intensity of Organizations in Knowledge Economy," Proceedings of the 3rd International Conference on Web Information Systems and Technologies, Barcelona, Spain, 210-213.

Čech, P. & Bureš, V. (2007). 'Recommendation of Web Resources for Academics - Architecture and Components,' Proceedings of the 3rd International Conference on Web Information Systems and Technologies, Barcelona, Spain, 437-440.

CISCO (2010). 'Planning the Migration of Enterprise Applications to the Cloud,' White Paper. [Online], [Retrieved April 4, 2012], http://www.cisco.com/.

Claypool, K. T., Jing, J. & Rundensteiner, E. A. (1998). "SERF: Schema Evolution through an Extensible, Reusable and Flexible Framework," Proceedings of the International Conference on Information and Knowledge Management, Worcester, MA, 1-8.

Coviello, A. et al. (2001). 'Standardized KM Implementation Approach,' IST Project No 2000-26393 Deliverable D 3.1. European KM Forum.

Davenport, T. H., Delong, D. W. & Beers, M. C. (1998). "Successful Knowledge Management Projects," *Sloan Management Review*, 39 (2), 43-57.

De Carvalho, R. B. & Ferreira, M. A. T. (2001). "Using Information Technology to Support Knowledge Conversion Processes," *Information Research*, 7 (1), paper 118.

EMC (2011). 'Planning a Data Center Migration: Five Key Success Factors,' *EMC Corporation*, Hopkinton, MA.

Firestone, J. M. (1998). "Basic Concepts of Knowledge Management," White Paper prepared for Executive Information Systems, Inc. Wilmington, DE.

FSA (2007). Data Migration Roadmap: A Best Practice Summary, Version 1.0. Department of Education Office of Federal Student Aid. [Online], [Retrieved February 24, 2012], http://federalstudentaid.ed.gov/static/gw/d ocs/ciolibrary/ECONOPS_Docs/DataMigratio nRoadmap.pdf.

Haller, K. (2009). "Towards the Industrialization of Data Migration: Concepts and Patterns for Standard Software Implementation Projects," *Lecture Notes in Computer Science*, 5565, 63-78.

Howard, P. (2011). 'Data Migration – 2011,' Bloor Research, London, UK.

Hubert, C. & O'Deil, C. (2000). "Successfully Implementing Knowledge Management," *American Productivity and Quality Center*. [Online], [Retrieved April 4, 2012], http://www.providersedge.com/docs/km_ar ticles/Successfully_Implementing_KM_- _APQC.pdf.

Hussain, F., Lucas, C. & Ali, M. A. (2004). "Managing Knowledge Effectively," *Journal of Knowledge Management Practice*, May 2004, 1-12.

Jorfi, S., Nor, K. M. & Najjar, L. (2011). "The Relationship between IT Flexibility, IT-Business Strategic Alignment, and IT Capability," *International Journal of Managing Information Technology,* 3 (1), 16 - 31.

Kantawala, A. (2008). "Case Study: Data Migration for a Global Semiconductors Manufacturer," Messner, W., Hendel, A. and Thun, F. (eds), *Rightshore!*: Successfully SAP® Project Ofshore, *Springer Verlag*, Heidelberg.

Manek, P. (2003). Microsoft CRM Data Migration Framework, White Paper. [Online],

[Retrieved April 4, 2012], http://download.microsoft.com.

Manion, J. (2011). "Streamlining the Reporting Process, Part 1," [Online], [Retrieved February 18, 2012], http://www.stratigent.com/community/web sight-newsletters/streamlining-reporting-process-part-1.

Mikulecký, P. (2003). "Information and knowledge Support of a Student in a University Environment," Proceedings of the International Conference on Computer as a Tool, Ljubljana, Slovenia, 108-111.

Patton, M. Q. (2001). "Evaluation, Knowledge Management, Best Practices, and High Quality Lessons Learned," *American Journal of Evaluation*, 22 (3), 329-336.

Pee, L. G. & Kankanhalli, A. (2009). "A Model of Organizational Knowledge Management Maturity Based on People, Process, and Technology," *Journal of Information & Knowledge Management*, 8 (2), 79-99.

PRINCE2 (2009). 'Managing Successful Projects with PRINCE2: Office of Government Commerce,' *The Stationery Office*, Norwich, UK.

QLOGIC (2008). "Data Migration – A Never-Ending Story," *White paper, QLOGIC*, Aliso Viejo, CA.

Semrádová, I. & Kacetl, J. (2011). "Ethics in the Future Manager's Professional Training," *E+M Ekonomie a Management,* 14 (2), 79-89.

SkySparc (2010). KEVA: Stremlining the Reporting process with Skyreport, SkySparc Wallstreet Excellence. [Online], [Retrieved February 18, 2012], http://www.skysparc.com/pdf/case_study_k eva.pdf.

Stenmark, D. (1999). "Using Intranet Agents to Capture Tacit Knowledge," *Proceedings of the WebNet 1999, Chesapeake,* Honolulu, Hawaii, 1000-1005.

Syntel (2006). Six Steps to Migration Project Success, *Applications: A White Paper Series, Syntel Inc., Troy, MI.*

Van der Spek, R. & Kingma, J. (1999). "Achieving Successful Knowledge Management Initiatives," CBI/IBM (eds), Liberating knowledge, business guide of Confederation of British Industry, *Caspian Publishing, London.*

Yelden, E. F. & Albers, J. A. (2004). 'The Business Case For Knowledge Management,' *Journal of Knowledge Management Practice*, August 2004, 1-12.

9

The Value Position of the Role of Knowledge Management and Its Benefits for Benchmarking Application

Barbora Jetmarová

University of Pardubice, Faculty of Economics and Administration, Pardubice, Czech Republic

Abstract

The purpose of this research is to describe the value position and benefits of the role of knowledge management in benchmarking application. The paper also seeks to participate in the current debate on developing the theoretical basis for benchmarking concept and benchmarking cycle. Benchmarking is a valuable management tool, which provides an opportunity to learn from other enterprises. Knowing how the best enterprises conduct their business is a critical element of successful enterprise. In today's world, successful enterprises are those that are innovative, flexible and are able to handle rapid change. This can be achieved by continuously learning from others, doing benchmarking studies to create new knowledge, adapting new best practices, and innovations, establishing a knowledge management infrastructure to capitalize on and disseminating results gained from benchmarking studies widely throughout the organization. On the one hand managing knowledge and effective knowledge management play an important role in successful benchmarking studies, on the other hand benchmarking has the potential for development of knowledge management in the enterprise. The paper finds that it is important to understand the systematic relationship between knowledge management and benchmarking and be aware of the value that can be generated in creating sustainable competitive advantage.

Keywords: benchmarking, benchmarking cycle, knowledge management.

Introduction

An economic environment is characterized by globalization of market, strong competition, technology advancement, increasing uncertainty and discontinuity. The World economy is getting more service driven and knowledge oriented. The main drivers of the worldwide change are innovations, which have been occurring at an unprecedented rate. The complexity of innovation has been increased by growth of knowledge, evolving technology, shorter product lifecycles and higher rate of new product development. (Du Plessis, 2007) Specially, technological developments and innovations in IT area have caused a

substantial increase in customer knowledge. Customers are now more educated and more demanding than ever. Harrington and Harrington (1995) state: "With intense competition in industry today, simply meeting or beating past performance will not result in the level of improvement necessary to remain competitive." In the fast changing business world of today, managing knowledge has become the basis of every organization.

Managers, owners and investors need to know the current situation of the company, business environment and competitors, in order to promote and maintain its position in the market. Appropriate knowledge allows them to take the right decisions when obtaining financial resources, in determining the optimal financial structure, in allocation of available funds, in the provision of trade credit, in the distribution of profits, etc. Adapting knowledge management itself is not enough. The current environment of globalization and economic turbulence has increased the challenges executives face and, therefore, the need to find the right tools to meet these challenges. Enterprises are constantly looking for new ways and methodologies to improve their performance and gain competitive advantage. Next to the knowledge management, enterprises have to consider and in many cases adapt or implement a wide range of innovative management philosophies, approaches, tools and techniques. To choose the right management tool, executives must be more knowledgeable than ever as they sort through the options and select the right management tools for their enterprises. Among the most popular management tools such as Strategic Planning, Mission and Vision Statements, Customer Relationship Management, Outsourcing, Balanced Scorecard etc. benchmarking has emerged as a useful, easily understood, and effective tool for remaining competitive. Many organizations already recognized the importance of benchmarking. According to Bain & Company, benchmarking is the most used management tool worldwide since 2009 (Rigby and Bilodeau, 2011).

Benchmarking

Benchmarking presents continuous, systematic monitoring and evaluation of how well and effectively the enterprise carry out the service or produce the product, as compared with enterprises that produce best practices. In case that the procedures are better elsewhere, the enterprise is trying to apply them to themselves, so that its efficiency match to the benchmark efficiency or even better is higher than the benchmark efficiency. Benchmarking is the process of comparing the enterprise with its competitors besides that, benchmarking is

also active in seeking the best ideas, methods and approaches, and generating knowledge that are applicable to the enterprise and could contribute to increase its efficiency. (Patton, 2001) If the benchmarking is done correctly, it is one of the most effective techniques for identifying and optimizing opportunities for implementing change and improving performance and thereby for increasing competitiveness.

Defining Benchmarking

Benchmarking as a management tool has many definitions. Robert. C. Camp, who stood at the birth of benchmarking, defines benchmarking as: "The search for organizations best practices that lead to superior performance". He also states: "Benchmarking is your key to become the best of the best." (Camp, 2006) Zairi (1994) defines benchmarking as continuous and systematic process of evaluating organizations recognized as leaders by their peers determining business and work process that represent best practices and establishing rational performance goals. Benchmarking has been variously defined and classified by different authors. As evidenced in literature, most authors have provided almost similar views on benchmarking. According to literature review, benchmarking definitions can be characterised into following major areas: measurement via comparison, identification of best practices, implementation, continuous improvement and systematic process in carrying out benchmarking activity. (Sarkis, 2001; Tölösi and Lajtha, 2000; Hodgetts et al., 1999; Ramabadron et al., 1997; Cooper et al., 1996; Voss et al., 1994; Anand and Kodali, 2008) Therefore, it is believed that these areas encompass pertinent aspects of any benchmarking process. After analysis of various benchmarking definitions, benchmarking can be described as: a management tool for attaining or exceeding the performance goals by learning from best practices and understanding the process by which they are achieved.

History and Types of Benchmarking

First comprehensive benchmarking project was carried by The Xerox Corporation. They used performance benchmarking for the first time in 1979. (Camp, 2006) Since that date, benchmarking procedures are constantly improving. Benchmarking is influenced by the development of management systems, statistical methods and information technology. In recent years, enterprises have recognized that focusing simply on performance measures and metrics leads to frustration because it is not clear as to how the leading performer achieves that performance. Nowadays, benchmarking is about improving competitive position, and using 'best practice' to stimulate radical innovation rather than seeking minor, incremental improvements on historic Performance. Benchmarking is a critical business process that is being continuously improved within most major organizations. (Zairi and Al-Mashari, 2005) Therefore, the term benchmarking describes numerous different activities. Several authors defined and classified many types of benchmarking.

The most commonly accepted types of benchmarking are described by Zairi and Al-Marshari (2005). They define two main types of benchmarking: performance and best practice benchmarking (process benchmarking). Performance benchmarking refers to comparison of process output as a means of identifying opportunities for improvement, setting performance targets and understanding relative positioning in comparison to other organizations. Best practice benchmarking refers to the comparison of the actual processes, practices and procedures in order to gain detailed knowledge of how improvements can be made. Both mentioned benchmarking types are closely linked and are widely used. While metric benchmarking answer the question "what" or "how many", process benchmarking seek the answer to "how" the organization achieves excellent performance. Metric benchmarking measure and compare the consequences, while process benchmarking look for the causes. Comparing the results is important to identify activities that need to improve, conversely comparing activities and processes shows activity that leads to better results. The process benchmarking can be internal, competitive, functional or generic. Global Benchmarking Network (GBN, 2011) asserts the same classification, but adds two main categories: informal and formal benchmarking. Informal benchmarking can be defined as an unstructured approach to learn from the experience of other organisations; therefore not following a defined process. Formal benchmarking is conducted consciously and systematically by organizations and is divided in two already mentioned categories: performance benchmarking and best practice benchmarking. Each benchmarking type has different requirements for knowledge management.

Knowledge Management

In the current decade knowledge as competitive asset is accepted universally and interest in knowledge management continues to grow. Benchmarking is very dependent on the availability of information and knowledge. To ensure successful benchmarking knowledge has to be identified and managed. Managing knowledge and effective knowledge management is very important for doing benchmarking studies. Key activities of knowledge management are creation of new knowledge, knowledge storage, knowledge distribution and knowledge application. Specially knowledge distribution is important for benchmarking. On the other hand, benchmarking is useful in creation of new knowledge.

Definition of Knowledge Management

Many Knowledge management definitions exist. For the purpose of this paper, only selected definitions will be focused on. Alavi and Leidner (1999) define knowledge management as a systemic and organizationally specified process for acquiring, organizing and communicating knowledge of employees so that others may make use of it to be more efficient and productive. According to Patton (2001) knowledge management is a collaborative management discipline that aims to make employees smarter, more innovative, and better decision makers. Daroch and McNaughton (2002) indicate that knowledge management is a management function that creates or locates knowledge, manages the flow of knowledge and ensures that knowledge is used effectively and efficiently for long-term benefit of the organization. Organization that demonstrates competence in knowledge management has knowledge-oriented and that knowledge management therefore becomes a guiding business philosophy that influences strategies undertaken by organizations managers. The ideal knowledge enterprise is the enterprise where people exchange knowledge across functional areas of the business by using technology and established processes. All the knowledge workers are in an environment where they can freely exchange and produce knowledge assets by using various technologies.

Knowledge

One of the sources for lasting competitive advantage is knowledge. Acquisition, exchange and creation of knowledge are crucial for benchmarking. Acquisition, creation and exchange of knowledge depends on the type of knowledge and takes places in various sources such as individual level, group level, and organizational level. Different types of knowledge can be learned and transferred by various ways. Lundvall and Johnson (1994) developed distinctions that are useful for understanding the different channels and mechanisms through which learning different types of knowledge takes place. Learning the four types of knowledge (know- what, know-why, know-how, know-who) tends to take place in different ways and through different channels. Important aspects of know-what and know-why may be obtained through

reading books, attending lectures and accessing data based. Know-how and know-who are more entrenched in practical experience. Written manuals may be helpful, but in order to use them some prior basic skills are needed. Know-how is typically learnt in apprenticeship-relationship where the apprentice follows the master. Know-how is what characterizes skilled worker and artisan but it is also something that distinguishes the first-rate from the average manager and scientist. Know-who is also learnt in social practice and some of munities of engineers and experts are kept together by reunions, conferences, professional societies, etc. giving the participants access to discussion of experiences and information battering with professional colleagues (Carter, 1989).

Not all types of knowledge can be transferred and learned, but they can be transformed into another form of knowledge. There is difference between explicit and implicit knowledge. Implicit knowledge is a knowledge which hasn't yet been codified but that can be codified. One way to make implicit knowledge explicit is to write it down. Knowledge that is written down may be passed on to others and be absorbed by those who can read and understand specific language. Even if the knowledge is written down, people need to have some prior knowledge about the topic, to understand the message. There is also tacit knowledge, which represents the valuable and highly subjective insights and intuitions that are difficult to capture and share because people carry them in their heads.

Difference is also between local and global knowledge. From the point of view of the whole economy, the transformation of local knowledge into global knowledge is of great interest. Education and training systems generalize knowledge and embody knowledge in people. One way to generalize knowledge is to codify it. Codification and efforts to make explicit what is implicit may be seen as one important way to enhance the capacity to share knowledge in society.

Another important mechanism for spreading experience-based knowledge is the mobility of workers. Other way to distribute knowledge is by benchmarking.

Benchmarking Cycle with Aspect to Knowledge Management

Benchmarking is the continuous learning process. In order to get useful results from benchmarking it is necessary to keep a systemic approach and respect benchmarking cycle. To initiate such cycle, management support is required. Over time, various methodologies were developed. Different sources describes phases and steps of benchmarking differently. The most important is the approach developed by four organizations, which are extensively involved in benchmarking (Boeing, Digital Equipment, Motorola and Xerox). This approach establishes the general context for the creation of a model, uses the four phases of benchmarking - planning, data collection, analysis and improvement through adaptation.

Figure 1 presents benchmarking model, which shows benchmarking cycle and was developed pursuant to wide literature research based on a comparison of several benchmarking models. It shows the benchmarking cycle and displays four phases of benchmarking - planning, data collection, analysis and adaptation. These phases are mutually intertwined. The left side of the picture shows what happens within own enterprise. The right side shows the steps which happens within the competitor's or benchmark's enterprise. More detailed description in Jetmarová, (2011).

Knowledge management has its use in all phases of benchmarking. The two firs phases: planning and data collection are phases, which benefits from knowledge management most. Knowledge management might be helpful in planning phase. Before starting a benchmarking programme, the enterprise must decide what to compare – what will be the subject of benchmarking. The subject should be chosen with reference to key criteria such as volume, cost, and value. If enterprise manages its knowledge, it is easier to set the subject of benchmarking as managers have good cognizance of different enterprises needs. Very important and sometimes problematic benchmarking phase is data collection. Some enterprise finds it difficult to get data from benchmarks. Collecting reliable data can be time consuming and requires thorough knowledge about own enterprise and also about benchmark as it is necessary to collect internal data in own enterprise and also external data from benchmark. Knowledge management helps greatly in internal data collection and is useful in external data collection. Knowledge management can facilitate collaboration with other enterprises and knowledge transfer across boundaries.

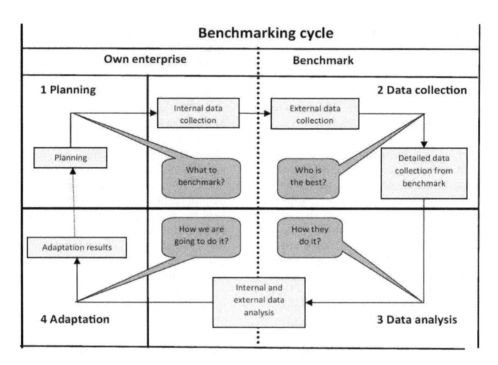

Fig 1. Benchmarking Cycle

Benefits of Benchmarking and Knowledge Management Co-Operation

Benchmarking and knowledge management are useful tools itself, but together they greatly benefit from each other. Benchmarking is extremely useful in developing knowledge management in organizations. On the other side, managing knowledge and effective knowledge management is very important for doing benchmarking studies.

The Value Position of the Role of Benchmarking in Knowledge Management

- One of the primary purposes of benchmarking is to evaluate, implement and spread information and knowledge. Benchmarking is useful in creation of new knowledge in enterprises. Best practices obtained by benchmarking have become the most sought out form of knowledge. (Not effective practices, or decent practices - but best.)

- Best practices and lessons learned due to benchmarking presents intellectual capital, which is meaningful form of knowledge. Acquiring knowledge and skills through benchmarking is efficient way of gathering better results and improving enterprises capacity without spending lot of time and money on developing innovations. Innovations - the main drivers of the worldwide change become more costly and risky due to pressure from market and fast technological changes.

- Knowledge management provides an overview of what is available in the enterprise. This allows enterprise understand which areas of knowledge are lacking and by benchmarking systematically gain the knowledge. In case that there are knowledge gaps in enterprise, benchmarking can fulfil them.

- Benchmarking, best practice research and leadership can be named as a knowledge management tools for gaining and sharing knowledge. Even if benchmarking is useful management tool for sharing knowledge, there are barriers influencing successful knowledge transfer. Main barriers are as follows: First biggest barrier to the transfer is unawareness on both ends of the transfer. Employees in most companies, neither the source nor the recipient know someone else have knowledge they require or would be interested in knowledge they have. Second biggest barrier to transfer is the absorptive capacity. Even if they know about better practice, they might not have the resources or enough practical detail to implement it. The third barrier is lack of pre-existing relationship between the source and recipient of knowledge. Fourth barrier is lack of motivation to transfer the knowledge. (O'del and Grayson, 1998)

- In many countries, numerous national initiatives continue to encourage organizations to benchmark, as they realize that current and future competition will be knowledge-based. Benchmarking is important aspect regarding to knowledge management. Transfer of best practice is the most common and most effective knowledge management strategy. Persistent to benchmarking company can develop a route map guiding the organization in the initial steps of implementing knowledge management. The instruction is derived from a study of successful knowledge management implementations in other enterprises that are successful in this area.

The Value Position of the Role of Knowledge Management in Benchmarking Application

- Benchmarking is using knowledge and experience of others to improve the enterprise. To ensure successful benchmarking knowledge has to be identified and managed. Therefore, benchmarking is very dependent on the availability of information and knowledge. Without knowledge management in organization, it is very difficult to carry out benchmarking.

- Knowledge management provides knowledge driven culture, which is helpful in benchmarking. Knowledge management helps to spread and transfer the useful practices around the enterprise acquired by benchmarking.

- Knowledge management assists in identifying gaps in the knowledge base and benchmarking provides information to fill them.

- Effective management of knowledge inside an organization is important for benchmarking application. Doing benchmarking and transferring best practices within an enterprise is much more effective, when it is part of an overall environment, which values the sharing of knowledge. O'del and Grayson (1998) state that when eneterprise does not have knowledge management, results from benchmarking will have only local benefit or will spread by luck.

- Knowledge management can facilitate collaboration with other enterprises and knowledge transfer across boundaries through ensuring that experts with relevant expert knowledge have opportunity to share their tacit knowledge through collaboration, which is needed for particular types of benchmarking. While doing external benchmarking enterprise have to work collaboratively across organizational boundaries to ensure sustained innovations and competitive advantage.

- Knowledge management assists in converting tacit knowledge to explicit knowledge. This adds a lot of value to the organization as it is known which knowledge is available, and it is retrievable for future re-use. (Du Plessis, 2007) Knowledge management also assists in generation tools, platforms and processes for tacit knowledge creation and sharing in organizations, which plays important role in benchmarking. One of the ambitions of knowledge management is to capture the tacit knowledge required by a business process and encourage knowledge workers to share and communicate knowledge with colleagues. With such knowledge, it is easier to determine which processes are more effective or less effective than others and helps to identify which processes should be benchmarked. Too often employees in one part of a business start from beginning on a project because the knowledge needed is somewhere else but not known to them.

Conclusion

Successful companies are those that consistently learn from others, do benchmarking studies to create new knowledge, disseminate it widely throughout the organization, quickly embody it in new technologies and products, establish a knowledge management infrastructure and adapt new best practices, in other words, enterprises that apply knowledge management and benchmarking. Based on this article, it is clear that knowledge management and benchmarking are useful itself. Both of them have great potential, but together they greatly benefit from each other. It is important to understand the systematic relationship between knowledge management and benchmarking and be aware of the value that can be generated in creating sustainable competitive advantage. It is clear that managing knowledge and effective knowledge management play an important role in successful benchmarking programme. In contrast, developing best practice and improving performance through benchmarking is a vital approach for sharing and transferring knowledge and developing knowledge management in the enterprise. Further research is required how the value of knowledge management can be maximized to ensure a more efficient and effective benchmarking results.

The article was supported by the "Student Grant Competition" at University of Pardubice - the project "Science and research activities supporting the program Economics and Management" - SGFES03/2012.

References

Alavi, M. & Leidner, D. E. (2001). "Knowledge Management and Knowledge Management Systems: Conceptual Foundations and Research Issues," *MIS Quarterly,* 25 (1), 107-136.

Anand, G. & Kodali, R. (2008). "Benchmarking the Benchmarking Models," *Benchmarking: An International Journal,* 15 (3), 257 - 291.

Camp, R. C. (2006). Benchmarking: The Search for Industry Best Practices That Lead to Superior Performance, *ASQ Quality Press,* Milwaukee.

Carter, A. P. (1989). "Know-How Trading as Economic Exchange," *Research Policy,* 18 (3), 155-163.

Cavusgil, S. T., Calantone, J. R. & Zhao, Y. (2003). "Tacit Knowledge Transfer and Firm Innovation Capability," *Journal of Business & Industrial Marketing,* 18 (1), 6-21.

Coopers, B. J., Lejny, P. & Mathews, C. M. H. (1996). "Benchmarking a Comparison of International Audit in Australia, Malaysia and Hong Kong," *Managerial Accounting Journal,* 11 (1), 23 –29.

Daroch, J. & Mcnaughton, R. (2002). "Examining the Link between Knowledge Management Practices and Innovation Performance," *Journal of Intellectual Capital,* 3 (3), 210-222.

Du Plessis, M. (2007). "The Role of Knowledge Management in Innovation," *Journal of Knowledge Managemet,* 11 (4), 20-29.

Global Benchmarking Network, (2011). Global Survey on Business Improvement and Benchmarking. [online], GBN [Retrieved 2011-08-20],

http://www.globalbenchmarking.org/image
s/stories/PDF/2010_gbn_survey_business_i
mprovement_and_benchmarking_web.pdf.

Harrington, H. J. & Harrington, J. S. (1996).
High Performance Benchmarking: 20 Steps to
Success, *Mcgraw-Hill,* New York.

Hodgetts, R. M., Kuratko, D. F. & Hornby, J. S.
(1999). "Quality Implementation in Small
Business: Perspectives from the Baldridge
Awards Winners," *SAM Advanced
Management Journal,* 64 (1), 37 - 47.

Jensen, M. B., Johnson, B., Lorenz, E. &
Lundvall, B. A. (2007). "Forms of Knowledge
and Modes of Innovation," *Research Policy,* 36
(5), 680-693.

Jetmarová, B. (2011). "Benchmarking –
Methods of Raising Company Efficiency by
Learning from the Best-in-Class," *E+M:
Ekonomics and Management,* 14 (1), 83-96.

Moriarty, J. P. & Smallman, C. (2009). "En
Route to a Theory of Benchmarking,"
Benchmarking: An International Journal, 16
(4), 484-503.

O'dell, C., Grayson, J. R., Jackson, C. &
Essaides, N. (1998). "If Only We Knew What
We Know: The Transfer of Internal
Knowledge and Best Practice," *Free Press,*
New York.

Patton, M. Q. (2001). "Evaluation, Knowledge
Management, Best Practices, and High
Quality Lessons Learned," *American Journal
of Evaluation,* 22 (3), 329-336.

Ramabadron, R., Dean, J. W. & Evans, J. R.
(1997). "Benchmarking and Project
Management: A Review and Organizational
Model," *Benchmarking: An International
Journal,* 4 (1), 47 - 58.

Rigby, D. & Bilodeau, B. (2008). Management
Tools & Trends 2011[online], *Bain &
Company* [Retrieved 2011-09-08],
http://www.bain.com/publications/articles/
management-tools-trends-2011.aspx.

Sarkis, J. (2001). "Benchmarking for Agility,"
Benchmarking: An International Journal, 8
(2), 88-107.

Tölösi, P. & Lajtha, G. (2000). "Toward
Improved Benchmarking Indicators,"
Telecommunication Policy, 24 (4), 347-357.

Voss, C. A., Åhlström, P. & Blackmon, K.
(1997). "Benchmarking and Operational
Performance: Some Empirical Results,"
*International Journal of Operations &
Production Management,* 17 (10), 1046 –
1058.

Zairi, M. (1994). Measuring Performance for
Business Results, *Chapman & Hall,* London.

Zairi, M. & Al-Mashari, M. (2005). "The Role
of Benchmarking in Best Practice
Management and Knowledge Sharing," *The
Journal of Computer Information Systems,* 45
(4), 14 -31.

10

Principal Starting Points of Organisational Knowledge Intensity Modelling

Tereza Otčenášková, Vladimír Bureš and Jaroslava Mikulecká

University of Hradec Králové, Hradec Králové, Czech Republic

Abstract

Current business environment is highly competitive. Nevertheless, the organisations have to manage their activities and operate under such uncertain and hard conditions. It remains important that this environment can be identified not only at the organisational level, but also at the national level, where particular countries compete, and at the sub-organisational level, in which individual departments struggle to get required resources. In order to succeed, the knowledge intensity plays a significant role at all these levels. The knowledge intensity can be defined as an extent in which the knowledge processes are performed and knowledge resources are utilised. Therefore, the knowledge intensity represents an indicator that is worthy to be monitored. This paper deals with the theoretical fundaments of this concept and outlines three potential approaches to knowledge intensity measurement. Additive model, multiplicative, and incremental models of knowledge intensity are introduced in particular sections of this paper. The main implication of the mentioned research is to present the possibility how to increase the organisational effectiveness and competitiveness. Both the limitations of the knowledge intensity modelling and further research options are also discussed.

Keywords: Competitiveness - Knowledge Intensity - Models of Knowledge Intensity.

Introduction

Obtaining and retaining the competitive advantage is the primary task of all subjects not only from the organisational perspective. As discussed within various sectors and industries, knowledge is considered to be one of the rare renewable resources (Davenport and Prusak, 1998, van Zolingen, Sreumer and Stooker, 2001), which moreover possesses a significantly substantial innovative potential and therefore can be further developed. It is necessary to measure and monitor the ability and willingness of particular subjects to effectively use knowledge, especially for the purposes of the comparison of their capabilities and market position. The aim of this paper is to establish theoretical fundaments of the knowledge intensity modelling which might represent a utilisable tool for the organisational evaluation and comparison in the realm of their competitiveness. Firstly, the paper determines the knowledge intensity concept and its context. In the next part three potential approaches to the knowledge intensity measurement are outlined. These are represented by the additive, multiplicative and incremental model of knowledge intensity. In the next section of this paper, both the limitations of the knowledge intensity modelling and further research options are mentioned and analysed. Finally, the discussed issues are concluded.

Knowledge Intensity Definition and its Context

The knowledge intensity measurement aims to provide another indicator of competitiveness monitoring not only at the organisational level, but also for the purposes of the comparison of entire sectors, national economics as well as supranational units. This indicator might also enable the identification of organisational potential and the areas ('gaps') for further improvement of the efficiency of the organisational and related knowledge processes. The knowledge intensity can be considered as a distinctive characteristic of the company department, organisation, a particular sector or the whole country perceived as a complex technical-economical-social system (Mildeová, 2005), and therefore should be modelled and monitored. As mentioned earlier, knowledge intensity might be measured at the organisational level. Chan argues that 'knowledge intensity increases with the rising complexity of business processes' (2009, 161). Moreover, Andreeva and Kianto (2011) prove its influence on the organisational innovation performance. Although the knowledge intensity is mentioned (Andreeva and Kianto, 2011, Makani and Marche, 2012), the particular and utilisable models are neither outlined nor discussed. Therefore, this paper focuses on the organisational perspective, because these issues are usually omitted and there are hardly any potential options of the quantification of the knowledge and related processes.

Autio (2000) defines knowledge intensity as the extent to which a firm depends on its knowledge as a source of competitive advantage. Davenport and Smith (2000) assert that knowledge-intensive companies will allocate more resources to knowledge management. Prashantham (2008) links the knowledge-intensive firms with the majority of workforce being highly qualified and engaged in knowledge work. He considers knowledge as inherent within the organisational knowledge-intensive activities. Willoughby and Galvin (2005) state that knowledge intensity positively relates to the extent of research and development activities, represents an internal source of innovations and determines the organisational ability to innovation

processes. Makani and Marche (2010) claim that the knowledge intensity comprises two critical dimensions - the worker and the organisation. Nevertheless, they emphasise that there is hardly any consensus on the definition of neither the knowledge intensity nor the knowledge-intensive organisation.

The aforementioned confirms particular findings and interest in the discussed issues. Nevertheless, the criteria with which would enable the classification of organisations according to their knowledge intensity are not described. For the purposes of this paper the knowledge intensity can be defined as an extent of the knowledge potential utilisation within the organisation.

Potential Possibilities of Knowledge Intensity Measurement

Various approaches and ways of perception to knowledge intensity are identifiable within the professional literature. The World Bank Knowledge Index or the Knowledge Economy index (Chen and Dahlman, 2005) are available for the comparison at the national level. At the individual level, the intensity of knowledge work (Holsapple, 2003) can be evaluated or the knowledge work intensity score can be assigned to particular tasks of an individual worker (Ramirez and Streudel, 2008). Nevertheless, the explicit description of ways how to measure the knowledge intensity at the organisational level has not been properly introduced yet. So far, only a general framework of knowledge intensity modelling was introduced (Bureš and Čech, 2007), but no particular models have been described. One of the reasons is the situational complexity and the necessity of considering the organisational context (Bureš, 2007). Among influential factors, the following can be mentioned: company size, financial results, organisational image, vision, strategy, structure, culture or the proficiency of management. The amount of the investments into science, research and education or whether the company is able to utilise the advantages of programmes supporting the innovations should be included among such aspects as well.

Knowledge Intensity Models

The knowledge intensity, similarly to other indicators and processes, can be modelled utilising various methods and approaches exemplified by the conceptual modelling (Otčenášková, Bureš and Čech, 2011) and formal or informal modelling (Mikulecký, 2010). The question - which approach offers the most accurate and the most practically utilisable result which would moreover enable the comparison of organisational development, various companies from a certain sector or different economic sectors among themselves in time - remains vague. In this paper three basic models are introduced. The first one is the additive model, the second one is the multiplicative model and the third one is the incremental model. These concepts are described in more detail below.

Additive Model of Knowledge Intensity

This model assumes that there are particular components within each organisation linked with knowledge and the extent of its utilisation considering the organisational potential. These components can fulfil the hypothetical maximal potential and therefore can be summed up. These parts can be represented by the essential organisational elements exemplified by the strategy, culture, process, structure, power&politics and information technology (Cao, Wiengarten and Humphreys, 2011) or by the technological infrastructure, organisational infrastructure, strategic leadership, organisational learning and knowledge culture (Šajeva, 2010). Within this paper, even more elements are utilised - these are employees, technologies, leadership style, and the like. The additive model represents the elementary model of knowledge intensity measurement and therefore supposes the highest level of simplification represented by the independence of the included components.

Therefore, the mechanistic approach can be employed to support this model construction. It grounds from the principle of reductionism and is based on the analysis of the whole system which is decomposed to basic elements which are further indivisible. The useful and supportive parallel classification of knowledge considering the level of detail can be used as well. Such concept was introduced by Wiig, de Hoog and van der Spek. They distinguish various knowledge spans - from knowledge domain, through sequential specification including knowledge region, knowledge section, knowledge segment, knowledge element and knowledge fragment to knowledge atoms (Wiig, de Hoog and van der Spek, 1997). The particular problems - in other words the mentioned atoms - can be consequently solved and dealt individually. When evaluated and summed up, these components practically represent the extent of the knowledge organisational intensity for the purposes of the additive model. If there is a need to increase the knowledge intensity of the organisation, the individual enhancement of a particular component would be sufficient without the reference to other elements or departments of the organisation.

From the theoretical perspective of the model, the ideal situation represents the presence of all components and their maximal utilisation within the organisation, possibly the maximal utilisation of the organisational potential. Regarding the provision of comparability, firstly it is crucial to convert all the particular components to the same measurement unit or dimension. For these purposes, the ideal way seems to be the percentage scale where 0 % represents the absence or absolutely ineffective utilisation of a particular component, and 100 % signifies maximal, mostly only hypothetical, in practice hardly achievable, utilisation of a particular component. The enumeration of the components possibly incorporated in the model aspirates to be relatively long whereas the above described components can be hierarchically decomposed through various levels to basic 'building blocks' which would enable to determine the extent of knowledge intensity (for example the classification to particular information, knowledge and communication technologies). Nevertheless, in practice the component might not even be present at all within the company. If the overall organisational potential is completely fulfilled by particular components, the company might be considered as 100 % knowledge-intensive no matter whether the organisation is more focused on employees and the organisational culture or whether it prioritises the processes and technologies.

Not only two organisations can be compared. The additive model might facilitate the comparison of particular departments within the organisation or the monitoring of various units. The described approach based on the additive model is general and signifies its usability nearly for any purposes of such monitoring. The fundamental blocks of the model can differ. Figure 1 illustrates this fact and exemplifies it while depicting the organisational and departmental modelling showing their potential basic elements. Obviously, the modelling can follow different perspectives. The appropriate one should be chosen according to the current conditions and demands of the monitoring.

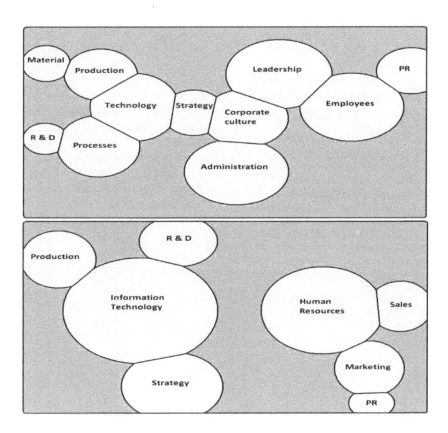

Fig 1. Additive Model: A - Organisational Components (Upper), B - Organisational Departments (Bottom) [Source: Author's Research]

To increase the knowledge intensity, a lot of activities and programmes can be employed. These might be exemplified by the establishment of knowledge manager position, the creation of the company university, quality circles, discussion forums and similar initiatives. The extent of the utilisation of these activities and their real (if any) contribution to the organisational knowledge intensity are not often necessarily unambiguous. The motivation of employees and other engaged subjects is undoubtedly another important part of the success of activities relating to the knowledge processes. Relative percentage share of particular components within the potential is therefore significantly different. Nevertheless, the definition of particular components itself remains arguable. The vague issues include for example the distinction of knowledge workers and knowledge processes or whether the extent of innovations within the company contributes to these processes. The determination of the most important component which should be represented to the higher extent is also controversial. This fact is naturally dependent not only on the internal factors and organisational characteristics, but also on the sector in which the company operates.

Multiplicative Model of Knowledge Intensity

The multiplicative model is based on the principles of multiplying particular elements of the system. Similarly to the additive and incremental model, the particular subsystems are 'put' together and their overall fulfilment and utilisation of the organisational potential is assessed.

In comparison with the additive model, this model assumes the interconnectedness of particular components and their dependence. Practically, this model seems to be more precise as well as relevant considering the usual reality within the companies/systems. Each element within the organisation or its department usually relates to one or more components. Some components are influenced by other ones and therefore overlapped (see Figure 2). Such overlays represent the connected areas which provide higher value added to the organisation, because the synergic effects emerge. There are parts included in more elements which increase the potential utilisation in more areas. As illustrated in Figure 2, the new technology implementation would positively influence more components than just itself. It would support the business strategy, improve organisational processes and enhance corporate culture.

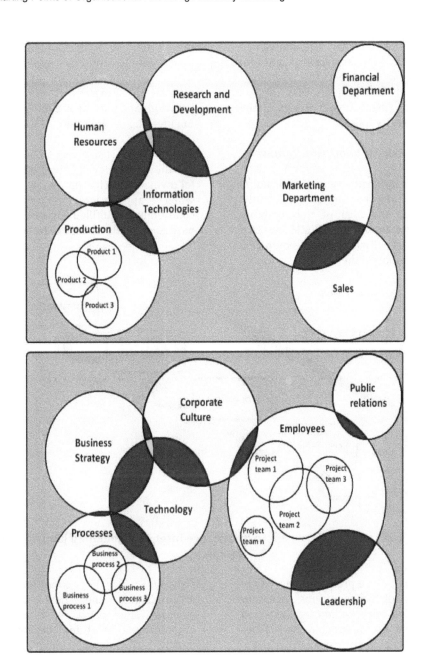

Fig 2. Multiplicative Model: A - Organisational Components (Upper), B - Organisational Departments (Bottom) [Source: Author's Research]

The elementary organisational elements are assessed for the purposes of both the additive and multiplicative models. Nevertheless, in the additive model these are summed up whereas the multiplicative model necessitates the multiplication of the overlapped components instead of the simple sum. The overall result of the extent of the organisational knowledge intensity is better - this means that the value of the knowledge intensity indicator is higher. The knowledge management can be considered as a useful 'tool' how to interconnect the particular components of the organisation or its departments to ensure higher values of the knowledge intensity index.

Incremental Model of Knowledge Intensity

The knowledge intensity incremental model differs from the aforementioned significantly. It is based on the idea of the Capability Maturity Model Integration (CMMI) where the maturity of the organisation is monitored. The implementation of the model is utilisable for the organisational efficiency improvement and its application would therefore represent an option how to monitor the progress of working with the knowledge within the organisation (Software Engineering Institute, 2011). Each organisation begins at the first, the least knowledge-intensive, level. The advancement to the next level, represented by the following maturity phase, is possible only when appropriate assumptions and conditions are fulfilled. Figure 3 illustrates all particular levels which can be achieved.

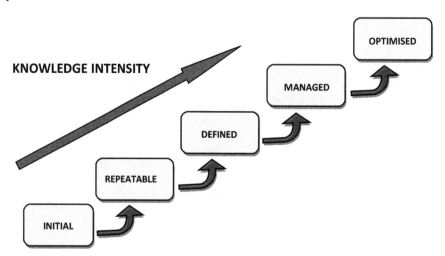

Fig 3. Application of CMMI Concept to Knowledge Intensity Modelling [Source: Author's Research]

The general - but more precise - overview of the description of the conditions necessary for the next level advancement is illustrated in Table 1 and the particular phases are introduced later. Nevertheless, the assumption of the enhancement is based on the principle that each level represents a particular added value what means that the previous level conditions remain accomplished and new improvements emerge. This paper is focused on the organisational level and therefore describes mostly the advancement from the perspective of knowledge processes because of their importance and the interconnectedness to other components.

Table 1: Conditions of Particular the Advancement to the Next Level [Source: Author's Research]

Particular Phase	Characteristics of Particular Components
INITIAL	
Leadership	No motivational schemes are used to support knowledge processes.
Corporate Culture	Corporate culture does not support knowledge processes, knowledge is not considered as an asset.
Strategy	Knowledge strategy is not available.
Processes	Processes are ineffective, separated and chaotic.
Technologies	Technologies hardly used, inappropriately chosen and implemented.
REPEATABLE	
Leadership	Knowledge aspects are considered in business strategy only.
Corporate Culture	Knowledge and knowledge processes are accepted to a limited extent only at the informal level.
Strategy	Only selected knowledge processes are supported, the approach is still neither systematic nor systemic.
Processes	Selected processes are monitored and modelled, knowledge aspects are omitted.
Technologies	Only basic information technologies are used, mostly only for communication purposes.
DEFINED	
Leadership	Active management of all defined knowledge processes occurs.
Corporate Culture	Formal description of supportive culture and motivational schemes emerge.
Strategy	Explicit form of knowledge strategy is available.
Processes	Processes with of knowledge inputs and outputs are fully documented and codified.
Technologies	Information and communication technologies are purposefully used for the support of knowledge processes.
MANAGED	
Leadership	Flexible management of knowledge processes and clear vision of further development occur.
Corporate Culture	Knowledge and knowledge processes are symbolised.
Strategy	Knowledge strategy is in alignment with general business strategy, however still separate.
Processes	The presence of formalised and managed business processes and their interrelation to knowledge processes is observable.
Technologies	Knowledge-based technologies and approaches are supported, partly implemented.
OPTIMISED	
Leadership	Management of knowledge processes based on knowledge strategy exists together with continual innovations.
Corporate Culture	Knowledge and knowledge processes become natural part of an organisational culture.
Strategy	Continuous improvement based on bottom-up and top-down approach is ensured, interwoven with general business strategy.
Processes	Continual development and alignment of business and knowledge processes with respect to corporate culture and business strategy are present and developed.
Technologies	Systemic implementation, development and utilisation of information, communication and knowledge technologies with alignment with general business strategy.

In the initial phase, the knowledge-oriented activities and process are chaotic, their management is not coordinated and the work with knowledge within the organisation remains neither systematic nor systemic. The knowledge utilisation emerges only on the urgency basis, knowledge is found at the moment when is necessary to quickly find the solution. However, hardly anybody knows who possesses which knowledge, who can be contacted in a certain situation, and so on.

The shift to the phase called repeatable is possible when the accomplishment of at least primary demands on activities and knowledge processes within the company is realised. At this level, these should be repeatedly applied. This means the possibility of the employment of the same - already proven - procedures and the avoidance of repeating the same mistakes and lapses experienced by the company in the past. This phase ensures to a limited extent the cost reduction of financial means used for both the searching of potential ways how to realise the activities and the effort to discover already known solutions of a particular problem.

The condition of the advancement to the phase defined is the provision of the option to determine, document and codify the particular processes. These are afterwards more easily applicable within various contexts. Moreover, the knowledge strategy is explicitly applied and shared within the organisation.

The level managed denotes the ability of the organisation to effectively control and flexibly apply knowledge during their management. The advancement to this phase requires the determination of knowledge strategy for the given company with particular linkage to business strategy. Potentially the knowledge manager might be formally appointed.

The organisation being at the highest level - the optimised one - is able to improve its processes continually and apply the innovations to them. From the technological perspective, the employment of Ambient Intelligence technologies at the workplace is realisable and would support the processes and correspondingly enhance the organisational knowledge intensity (Mikulecký, 2007). Moreover, the strategic vision and knowledge strategy are effectively interconnected with the business strategy and support it. Therefore, significant cost reduction and competitiveness improvement occur.

The implementation of CMMI and its application in practice should facilitate not only the efficiency and effectiveness enhancement of the organisational activities, but also the quality improvement.

Limitations and Further Research

The general methodology can be considered as an initial overview of potential options how to measure the knowledge intensity. Nevertheless, the organisational context with its specifics certainly necessitates to be taken into account. Moreover, the discussed issues are complex as well as dynamic and uncertain. Therefore, this realm generally evokes more areas for further research.

Firstly, the current monitoring of 'knowledge intensity' within organizations together with their attitude to issues linked with knowledge should be researched and evaluated to recognise which areas compel the most attention. This process would afterwards enable better resource allocation and more appropriate investments. Not only the current state should be mapped, but the results are usable for the long-term sustainability of the organisational prosperity.

Secondly, the methodology itself should be tested in practice. For example, a case study relating to these issues would be useful for both the practical usability and the identification of the model weaknesses. The potential   areas   for   appropriate improvements will be revealed and aptly amended or included.

Moreover, the fact that a lot of particular input parameters are qualitative complicates the indicator determination. Subjectivity of their perception and definition influences their identification and measurement. Therefore, the standardisation and optimisation would contribute to the elimination of these unfavourable phenomena and to the increase of indicator comparability.

Furthermore, the formalisation of the models and indicator determination represents not only a difficult challenge, but nearly a need for its further utilisation and wider applicability. The formalised and more precise results would support the organisational competitiveness enhancement. On the contrary, the employment of all the mentioned knowledge intensity measurement improvements is connected with higher demands on financial, time, human and technical resources during the measurement process and during the identification of potential gaps for advances. This would consequently make this process more problematic and less realisable for organisations.

At the moment, some potential options of knowledge intensity modelling and measurement are outlined. Nevertheless, these concepts require to be further elaborated and their usability afterwards verified in praxis within certain organisations through their various types, sectors, category of economic activity, main innovation focus, and the like. This process should ensure the increase of knowledge intensity indicator relevancy together with its higher applicability. Moreover, it should provide the determination of weights of particular components from the perspective of various subjects operating within different spheres. Possibly, the appropriateness of various approaches and models will be proved for the purposes of various sectors, companies and the like. These differences

might be revealed while examining the models practically.

Conclusion

Currently, the tool utilisable for the measurement of knowledge and organisational potential utilisation at the organisational level is not available. Therefore, this paper deals with the introduction of theoretical fundaments of a knowledge intensity concept usable for such purposes - the evaluation of organisational weaknesses, the identification of areas of further improvement and development and the consequent increase of its competitiveness. Three models of knowledge intensity modelling are described. These include the additive, multiplicative and the incremental one. Nevertheless, it remains important to up-to-date and amend the mentioned models and approaches according to the topical situational development and to changing internal and external conditions from both perspectives - the practical and theoretical one.

Acknowledgement

This paper was written with the support of specific research project 'The Research of the Ambient Intelligence Technologies Impact on the Intellectual Capital Development' which is a part of a GAČR project SMEW - Smart Environments at Workplaces No. 403/10/1310 and the project 'Innovation and support of doctoral study program (INDOP)' No. CZ.1.07/2.2.00/28.0327 financed from EU and Czech Republic funds.

References

Andreeva, T. & Kianto, A. (2011). "Knowledge Processes, Knowledge-Intensity and Innovation: A Moderated Mediation Analysis," *Journal of Knowledge Management,* 15(6), 1016-1034.

Autio, E., Sapienza, H. J. & Almeida, J. G. (2000). "Effects of Age at Entry, Knowledge Intensity, and Imitability on International

Growth," *Academy of Management Journal,* October 2000.

Bureš, V. & Čech, P. (2007). "Complexity of Ambient Intelligence in Managerial Work," ITICSE 2007: 12th Annual Conference on Innovation & Technology in Computer Science Education, Dundee (Scotland): University of Dundee, 325-325.

Bureš, V. & Čech, P. (2007). 'Knowledge Intensity of Organizations in Knowledge Economy,' INSTICC: 3rd International Conference on Web Information Systems and Technologies, Barcelona (Spain), 210-213.

Cao, G., Wiengarten, F. & Humphreys, P. (2011). "Towards a Contingency Resource-Based View of IT Business Value," *Systemic Practice & Action Research,* 24(1), 85-106.

Chan, J. O. (2009). "A Conceptual Framework for an Integrated Knowledge-Driven Enterprise Model," *Journal of International Technology and Information Management,* 18(2), 161-185.

Chen, D. H. C. & Dahlman, C. J. (2005). "The Knowledge Economy, the KAM Methodology and World Bank Operations," [Online], *The World Bank,* [Retrieved November 16, 2011], Available: http://siteresources.worldbank.org/KFDLP/Resources/KAM_Paper_WP.pdf.

Davenport, T. H. & Prusak, L. (1998). Working Knowledge. 1st edition, *Boston: Harvard Business School Press,* ISBN: 0875846556.

Davenport, T. H. & Smith, D. E. (2000). "Managing Knowledge in Professional Service Firms," *The Knowledge Management Yearbook 2000-2001,* Oxford: Butterworth-Heinemann.

Holsapple, C. W. (ed.) (2003). Handbook on Knowledge Management: Knowledge Matters. 1st edition, Berlin - *Heidelberg: Springer-Verlag,* ISBN: 3-540-43527-1.

Makani, J. & Marche, S. (2010). "Towards a Typology of Knowledge-Intensive Organizations: Determinant Factors," *Knowledge Management Research & Practice,* 8(3), 265-277.

Makani, J. & Marche, S. (2012). "Classifying Organizations by Knowledge Intensity - Necessary Next-steps," *Journal of Knowledge Management,* 16(2), 243-266.

Mikulecký, P. (2007). 'Ambient Intelligence in Decision Support,' 7th International Conference on Strategic Management and its Support by Information Systems, Celadna (Czech Republic), 48-58.

Mikulecký, P. (2010). "Possibilities for Formal Models of Smart Environments," DNCOCO 10: 9th WSEAS International Conference on Data Networks, Communications, Computers, University Algarve, Faro (Portugal). In: Advances in Data Networks, Communications, Computers, Book Series: Advances in Data Networks Communications Computers-Proceedings, Mastorakis N. E. and Mladenov, V. (eds.), 131-136.

Mildeová, S. (2005). 'The Principles of System Dynamics Towards Balanced Scorecard Implementation,' 13th Conference on Interdisciplinary Information Management Talks, Budweis (Czech Republic): Schriftenreihe Informatik, Book Series, 16, 119-127.

Otčenášková, T., Bureš, V. & Čech, P. (2011). 'Conceptual Modeling for Management of Public Health in Case of Emergency Situations,' International Conference on Knowledge Engineering and Ontology Development, Paris (France), 344-348.

Prashantham, S. (2008). The Internationalization of Small Firms: A Strategic Entrepreneurship Perspective. *USA, Canada: Routledge.*

Ramirez, Y. W. & Streudel, H. J. (2008). "Measuring Knowledge Work: The

Knowledge Work Quantification Framework," *Journal of Intellectual Capital,* 9(4), 564-584.

Software Engineering Institute. (2011). "Capability Maturity Model Integration," [Online], *Carnegie Mellon,* [Retrieved November 25, 2011], Available: http://www.sei.cmu.edu/cmmi/.

Šajeva, S. (2010). "The Analysis of Key Elements of Socio-Technical Knowledge Management System," *Economics & Management,* 765-774.

Van Zolingen, S. J., Streumer, J. N. & Stooker, M. (2001). "Problems in Knowledge Management: A Case Study of a Knowledge-Intensive Company," *International Journal of Training & Development,* 5(3), 8-185.

Wiig, K. M., de Hoog, R. & van der Spek, R. (1997). "Supporting Knowledge Management: A Selection of Methods and Techniques," *Expert Systems with Applications,* 13(1), 15-27.

Willoughby, K. & Galvin, P. (2005). "Inter-Organizational Collaboration, Knowledge Intensity, and the Sources of Innovation in the Bioscience-Technology Industries," *Knowledge, Technology, & Policy,* 18(3), 56-73.

11

Enterprise 2.0 Technologies for Knowledge Management: Exploring Cultural, Organizational & Technological Factors

Umar Ruhi[1] and Dina Al-Mohsen[2]

[1]Telfer School of Management, University of Ottawa, Ottawa, Canada

[2]E-Business Technologies, University of Ottawa, Ottawa, Canada

Correspondence should be addressed to: Umar Ruhi; Umar.Ruhi@uottawa.ca

Academic Editor: Rene Leveaux

Abstract

This paper reports findings from a recent empirical study conducted to explore sociological and technological factors that affect the use of enterprise 2.0 (E2.0) technologies for knowledge management (KM). To help organizations adopt and institutionalize effective KM strategies, this study aims to highlight the effects of national and organizational cultural differences among operating environments of different firms, and to identify how these differences translate into varying knowledge management behaviors and use of E2.0 technologies for KM in firms.The study utilized a quantitative empirical research design to collect and analyze quantitative data from employees of various organizations in different countries and industries. A web-based survey data was collected from various countries including Canada, USA, and Saudi Arabia. Exploratory factor analysis and structural equation modeling techniques were used to estimate a structural model among factors impacting the use of E2.0 technologies for KM.The key findings from this study validate the role of technology perceptions including ease of use, usefulness, media richness and technology sophistication in improving the use of enterprise 2.0 technologies in the workplace. Furthermore, the use of these technologies was shown to have a positive effect on the knowledge management environment of the organization. In terms of cultural differences, the knowledge management environment of firms was shown to be affected by long-term orientation of the national culture. This study offers recommendations for companies operating in global cultural contexts on how to approach KM strategies differently according to national culture and organizational environments of firms.

Keywords

Enterprise 2.0, Knowledge Management, Technology Acceptance, National Culture, Organizational Culture, Corporate Strategy

Introduction

Organizational knowledge is considered an important determinant of profitability and a key driver for strategy development, value creation, and market competitiveness (Nonaka & Takeuchi, 1995; Marquardt & Kearsley, 1999; He & Wei, 2009). Consequently, effective knowledge management has been linked to overall corporate prosperity and superior business performance (Davenport & Prusak, 2000; Riege, 2005; Velev & Zlateva, 2012). This can be achieved when organizations harness innovation and creativity by incorporating knowledge about employee experience and relations into business-specific processes (Velev & Zlateva, 2012). On a global level, organizations worldwide are becoming increasingly aware of the benefits that can be achieved by engaging in effective knowledge management practices. However, additional research is needed to understand how organizational settings and cultural differences can translate into varying needs and outcomes for knowledge management practices, processes and technologies. Toward this, many researchers seek to further investigate the determinants of effective knowledge management by considering different organizational cultural contexts and industrial settings (Kim & Lee, 2006; Detlor et al., 2006; Alhamoudi, 2010). The purpose of this article is to contribute to this body of knowledge and help gain a better understanding of the cultural, organizational and technological factors affecting knowledge management behavior across multiple national and corporate settings.

Conceptual Foundation

The main research question under investigation in this study is as follows:

What are the effects of national cultural traits, organizational knowledge management context, and enterprise 2.0 technology attributes on the knowledge management practices of organizations, and consequently, the use of Enterprise 2.0 technology for knowledge management?

The conceptual framework shown in Figure 1 below depicts the overall orientation of this study and the core ideas that underpin the research investigation. The components of the conceptual framework are outlined in this section.

Adopting Davenport & Prusak's perspective of knowledge, we describe it as actionable information that is the outcome of individual or organizational experiences and values, combined with other relevant contextual information (Davenport & Prusak, 2000).

Additionally, for this paper, knowledge management (KM) is defined as the systematic, effective management and utilization of an organization's knowledge resources and encompasses the creation, storage, arrangement, retrieval, and distribution of an organization's knowledge (Saffady, 1998).

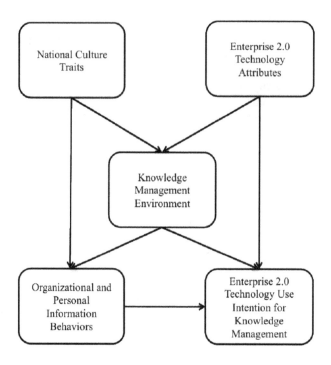

Figure 1: Conceptual Framework for This Study

The extant literature has shown that effective KM strategies and practices are predicated upon proper alignment with organization's knowledge management context (Detlor et al., 2006). This paper pertains to various sociological and technological factors related to the knowledge management context of an organization, and categorizes these factors as macro- or micro- context as discussed herewith.

In this paper, the macro context for KM is conceptualized using the notion of national culture. Research shows that there is a strong relationship between knowledge management and culture in that national cultural differences affect the process of knowledge management in many ways (Voelpel & Han, 2005). As such, tools and practices suitable for one culture may not work well in another (Holden, 2002; Stankosky, 2005; Pauleen, 2007; Pawlowski & Bick, 2012). As stated by Hofstede (1980; 1997) in his seminal work on cultural

dimensions theory, "there are no universal solutions to organization and management problems", and this certainly applies to institutionalizing knowledge management practices differently across national cultures.

At a micro-level, this study explores the knowledge management setup in an organization through the lens of knowledge management environment which represents "the culture and commitment within the organization to implement and institutionalize effective information and knowledge sharing processes, practices and technologies" (Detlor et al., 2006). A conducive knowledge management environment is seen as a key enabler that positively influences the effectiveness of KM practices (Holsapple & Joshi, 2000). Along with KME, we also explore two other related concepts – Organizational information behaviors that pertain to the information and knowledge sharing practices at the corporate level, and personal information behaviors that concern an employee's own actions in carrying out information and knowledge sharing practices (Detlor et al., 2006). Together, these constructs aim to capture the differences among various organizational knowledge contexts.

In this paper, the sociological factors outlined above will be considered in tandem with technological factors that are posited to impact the adoption of enterprise 2.0 (E2.0) technologies in the organization. Throughout the extant literature, there is a common theme that technology is considered to be a strong and effective enabler of KM best practices (Alavi & Leidner, 2001; Davenport & Prusak, 2000; Richter et al., 2013). However, traditional knowledge management systems (KMS) have been known to be rigid in their deployment and use, and suffer from many limitations (Richter et al., 2013; Ruhi & Choi, 2013), and organizations today need interactive KM technologies in order to foster social processes core to knowledge management practices (Ardichvili et al., 2003). This may be made possible through the use of E2.0 technologies (Richter et al., 2013).

In this paper, the term Enterprise 2.0 is used to refer to social computing platforms and channels (e.g. blogs, wikis, social networks, multimedia sharing sites) that can be used in organizations to support collaborative knowledge work. McAfee (2006) defines Enterprise 2.0 as the strategic integration of Web 2.0 technologies into an organization's intranet, extranet, and business processes. In the preceding definition, Web 2.0 refers to web applications that facilitate content creation, information sharing, interoperability, user-centered design, and enhanced collaboration (O'Reilly, 2007). In a KM context, such online interactive social applications are promising as they can help enable people to connect, communicate, collaborate and share information more actively than what was possible in traditional KMS (Kane et al., 2009; Zheng et. al, 2010). The use of such technologies for KM also marks a technology shift from a centralized knowledge-warehouse approach to a more dynamic, communication-based or network approach (Kuhlen, 2003; Hazlett et al., 2005).

Theoretical Model

To proceed with the investigation of sociological and technological factors affecting the use of enterprise 2.0 technologies for knowledge management, a theoretical model was formulated to test and validate various relationships among constructs related to national cultural traits, enterprise 2.0 technology attributes, organizational and personal information behaviors, knowledge management environment, and the use of enterprise 2.0 technologies for knowledge management. Figure 2 depicts the theoretical model for this research study.

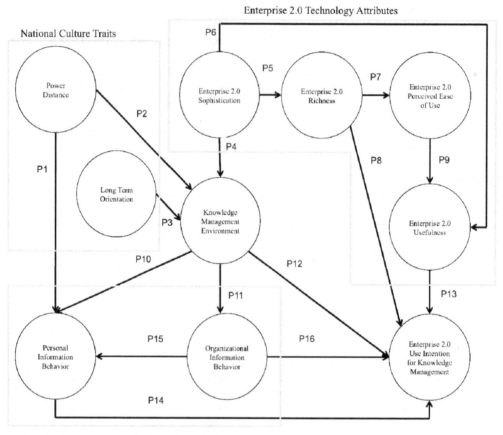

Figure 2: Theoretical Model for This Study

In formulating the theoretical model, this research draws upon five key theoretical frameworks from the extant literature: Hofstede's National Culture Dimensions (Hofstede, 1980; 1997), Davis et al.'s Technology Acceptance Model (TAM) (Davis, 1989; Davis et al., 1989), Daft and Lengel's Media Richness Theory (Daft & Lengel, 1986), DeLone and McLean's Information Systems Success Model (DeLone & McLean, 1992; DeLone & McLean, 2003), and Detlor et al.'s Knowledge management Context model (Detlor et al., 2006). Table 1 summarizes the various themes and constructs in the theoretical model and provides a brief description for each construct in the model. The constructs and interrelated propositions are described below.

Table 1: Theoretical Model Components

Theme	Constructs	Definition & Theoretical Basis
National Culture Traits	Power Distance	The extent to which members of an organization in a specific culture accept and expect inequality in the distribution of power (Hofstede, 1980)
	Long Term Orientation	Long Term Orientation is associated with values of perseverance and future planning. It does not have a high respect for tradition, fulfilling social obligations, and protecting one's 'face' (Hofstede, 1997).
Enterprise 2.0 Attributes	Perceived Ease of Use	Employee's expectation of the targeted system's required level of effort (Davis & Bagozzi, 1989).
	Perceived Usefulness	The degree of belief that using a particular system will enhance an employee's job performance (Davis & Bagozzi, 1989).
	Enterprise 2.0 Richness	Demonstrating the richest available medium of communication to convey messages properly and to ensure successful communication (Daft & Lengel, 1986).
	Enterprise 2.0 Sophistication	Tools diversity and maturity to enhance the end-user's technology interaction and overall use (Ghobakhloo et al., 2011).
Knowledge Management Context	Knowledge Management Environment	Context and culture of an organization that nurtures a knowledge management initiative (Detlor et al., 2006).
	Personal Information Behavior	Individual's own actions and practices in exchanging information and collaborating with others (Detlor et al., 2006)
	Organizational Information Behavior	The practices that employees observe and draw upon for information and knowledge sharing at the organizational level (Detlor et al., 2006).
Enterprise 2.0 Use Intention for Knowledge Management	Intention of Use of System	Intention to continue using Enterprise 2.0 for knowledge management (Davis, 1989; Davis & Bagozzi, 1989; DeLone & McLean, 1992; 2003).

Firstly, with respect to national cultural traits, the theoretical model in this study posits espoused national cultural values as individual difference variables that affect the knowledge management environment and practices in the organization. We draw upon Hofstede's National Culture Dimensions (Hofstede, 1980; 1997) and consider the two dimensions related to power distance and long-term orientation as parameters to differentiate cultures and to measure how

such values impact the workplace and employees attitudes (Boyd, 2012).

Propositions pertaining to national cultural traits are as follows:

- *P1: Greater Power Distance among an organization's employees has a positive impact on their Personal Information Management Behavior*
- *P2: Greater Power Distance among an organization's employees has a*

positive impact on the Organization's Knowledge management Environment

- *P3: Long-Term Orientation has a positive impact on the Organization's Knowledge management Environment*

Secondly, in terms of constructs pertaining to technology attributes of enterprise 2.0, the theoretical model includes factors that are expected to impact the employee's cognitive, affective and behavioral responses towards enterprise 2.0 technologies. The model draws upon the Technology Acceptance Model (TAM) (Davis, 1989; Davis et al., 1989) to conceptualize user acceptance of enterprise 2.0 technologies based on their perceptions of usefulness and ease of use. Additionally, media richness of E2.0 technologies was considered to be an essential determinant of adoption of these technologies. Media Richness Theory posits that the richest medium of communication is one that promotes understanding in a timely fashion; meanwhile, a channel that takes more time to understand is a less rich medium (Daft & Lengel, 1986). In determining the richness and usefulness of a medium, this study considers the sophistication of enterprise 2.0 tools in terms of the breadth of features and functions that are available to the end-user through these tools.

Propositions pertaining to technological constructs in the theoretical model are as follows:

- *P4: Sophistication of Enterprise 2.0 technologies has a positive impact on the Organization's Knowledge Management Environment*
- *P5: Sophistication of Enterprise 2.0 technologies has a positive impact on the Richness of Enterprise 2.0 technologies*
- *P6: Sophistication of Enterprise 2.0 technologies has a positive impact on the Perceived Usefulness of Enterprise 2.0 technologies*

- *P7: Richness of Enterprise 2.0 technologies has a positive impact on the Perceived Ease of Use of Enterprise 2.0 technologies*
- *P9: Perceived Ease of Use of Enterprise 2.0 technologies has a positive impact on the Perceived Usefulness of Enterprise 2.0 technologies*

The third grouping of constructs in the theoretical model pertains to the overall organizational knowledge management context, and it consists of the knowledge management environment, personal information behaviors, and organizational information behaviors. The operationalization of these constructs and their interrelationships is adopted primarily from Detlor et al. (2006) as outlined in the following propositions:

- *P10: Organization's Knowledge Management Environment has a positive impact on Personal Information Management Behavior*
- *P11: Organization's Knowledge management Environment has a positive impact on Organizational Information Behavior*
- *P15: Organizational Information Management Behavior has a positive impact on Personal Information Behavior*

Finally, intention to use enterprise 2.0 technologies for KM is positioned as the ultimate consequent variable in the theoretical model. This is aligned with other theoretical models in the information systems literature including the technology acceptance model (TAM) (Davis, 1989; Davis & Bagozzi, 1989) and the IS success model (DeLone & McLean, 1992; DeLone & McLean, 2003). The extant research literature also shows the context in which the knowledge is exchanged affects employee behavior in terms of creating, transferring and sharing knowledge through appropriate technologies (Malone, 2003).

The various posited relationships between other theoretical constructs and the intention to use E2.0 technologies for KM are as follows:

- *P8: Richness of Enterprise 2.0 technologies has a positive impact on the Intention to Use Enterprise 2.0 technologies for Knowledge Management*
- *P12: Organization's Knowledge Management Environment has a positive impact on the Intention to Use Enterprise 2.0 technologies for Knowledge Management*
- *P13: Perceived Usefulness of Enterprise 2.0 technologies has a positive impact on the Intention to Use Enterprise 2.0 technologies for Knowledge Management*
- *P14: Personal Information Behavior has a positive impact on the Intention to Use Enterprise 2.0 technologies for Knowledge Management*
- *P16: Organizational Information Behavior has a positive impact on the Intention to Use Enterprise 2.0 technologies for Knowledge Management*

Methodology

Data for this study were collected through an online survey that was administered to employees of various organizations across three countries – Canada, USA and Saudi Arabia. For respondents in Saudi Arabia, the survey questionnaire was accessible in both English and Arabic. Call for participation in the research study was communicated through personal contact with key personnel in various organizations, and it was also posted on various open and closed online social networks. The sampling techniques used were primarily convenience and self-selection based. Qualifying questions were asked at the beginning of the survey to ensure that respondents had some familiarity with social media tools for knowledge work in an organizational context.

The survey comprised demographic information questions, technographic behavioral items, questions about work atmosphere, and psychographic perceptions based questions. While the demographic and technographic sections of the survey were operationalized through direct questions consisting of an inventory of possible responses, the questions pertaining to constructs in the theoretical model comprising latent variables were operationalized using psychometric scales with responses on a Likert-scale and through categorical response type questions. Furthermore, most survey questions pertaining to the theoretical constructs were adapted from item scales that had been previously used and validated in other research. Appendix A lists the survey measurement items that were utilized for each of the constructs in the theoretical model.

Prior to rollout, a draft version of the survey was pre-tested using 15 respondents from the planned sampling frame, allowing the language of the survey to be improved, words to be better translated and item measurement scales to be verified. Pre-testing of a survey instrument is regarded as a crucial step in the development and design of a survey questionnaire ensuring the adequacy of planned survey administration and data collection procedures (Andrews et al., 2001).

In terms of analysis procedures, the demographic and technographic variables were analyzed using descriptive statistics and non-parametric statistical tests. Testing of theoretical constructs and relational propositions from the theoretical model was conducted through exploratory factor analysis and structural equation modeling techniques.

To examine the measurement model, the two-step approach suggested by Anderson and Gerbing (1988) was utilized whereby examination of the measurement model was conducted before testing the structural model. Both the measurement and structural models were estimated by using the structural equation modeling facilities of Smart PLS (Ringle et al., 2013). The PLS approach was chosen since it fits small-sample exploratory research (Gefen et al., 2000) and it does not require meeting the multivariate normality assumptions posed by other structural equation modeling techniques (Thomas et al., 2005).

Harman's post hoc one-factor test (Podsakoff et al., 2003) was conducted to examine common method bias in the data. Principal component factor analysis (unrotated solution) revealed that nine factors were extracted and the first factor accounted for 26% of the variance. Hence, common method bias was not a serious problem with the data because multiple factors emerged and no single factor accounted for a majority of the variance (Podsakoff and Organ, 1986; Podsakoff et al., 2003).

Results

A total of 350 responses were collected from across the three countries. After discarding partial responses, a total of 85 responses from Saudi Arabia and a total of 91 responses from North America (43 from Canada and 48 from USA) were retained for statistical analysis. Statistical analysis was conducted separately on these two datasets from Saudi Arabia and North America.

With respect to demographics, the average age of respondents fell around the 30–39 years frequency category for both samples. Males comprised 52% of the sample from Saudi Arabia, and 58% of the sample from North America. In other words, the majority of respondents were male. 42% of respondents from Saudi Arabia had undergraduate degrees while 48% of North American respondents had graduate degrees.

Majority of the respondents from both samples reported working in middle management back-office positions in large organizations (firms with more than 1000 employees).

With respect to technographic patterns in the behavior of using enterprise 2.0 technologies, most Saudi Arabian respondents reported using these technologies primarily for internal communication purposes within the organization. In contrast, respondents from Canada and the US reported internal collaboration being the most common use for enterprise 2.0 technologies. Furthermore, the results indicated that most Saudi Arabian employees use these tools for personal communication purposes rather than work related purposes. This pattern can possibly be related to another finding that suggested that companies in Saudi Arabia lagged behind those in Canada and the USA in terms of clearly defined guidelines and policies for the use of social media tools at the workplace.

The measurement model was assessed by combining the data from the two samples. Tables 2, 3 and 4 consist of matrices showing various tests for discriminate and convergent validities for the constructs in the theoretical model.

First, the loadings and cross-loadings of indicators were examined as a basic test for discriminant validity of measures. The matrix of loadings and cross-loadings is presented in Table 2 with the highest loadings of items shown in bold. In order to ascertain discriminant validity, the loadings of an item on its associated latent construct (target variable) should be higher in comparison to the item's cross-loadings on other latent constructs. All model constructs satisfied these criteria.

Table 2: Matrix of Loading and Cross Loadings

Measurement Items	Model Constructs									
	EOU	KME	LTO	OIB	PD	PIB	Richn	Soph	UI	Usefl
E2.0UI1	0.5676	0.2063	0.2880	0.1392	0.3872	0.2727	0.6326	0.4293	**0.9255**	0.5266
E2.0UI2	0.5021	0.1921	0.2252	0.1703	0.3422	0.2751	0.5350	0.3712	**0.9035**	0.5094
E2.0UI3	0.4826	0.2296	0.3163	0.1579	0.3429	0.3285	0.5512	0.3305	**0.8552**	0.4731
E2.0UI4	0.4778	0.2297	0.3013	0.2227	0.3646	0.2641	0.5173	0.3378	**0.8913**	0.5295
E2.0EOU1	**0.8654**	0.2386	0.3807	0.1320	0.3622	0.2738	0.6176	0.3825	0.4875	0.5296
E2.0EOU2	**0.9393**	0.2896	0.3859	0.1559	0.4212	0.3240	0.5340	0.3215	0.5068	0.5050
E2.0EOU3	**0.9038**	0.2454	0.4193	0.1596	0.4644	0.3775	0.5454	0.3458	0.5445	0.6078
E2.0R1	0.5505	0.2509	0.2743	0.2096	0.2622	0.4287	**0.7877**	0.4462	0.5489	0.4677
E2.0R2	0.4472	0.1919	0.2277	0.1269	0.1687	0.2106	**0.8087**	0.3429	0.4121	0.3704
E2.0R3	0.4900	0.2079	0.2116	0.2538	0.1536	0.3085	**0.8926**	0.4081	0.4756	0.3691
E2.0R4	0.5981	0.2037	0.2837	0.1819	0.3239	0.3407	**0.8680**	0.3576	0.6336	0.4466
E2.0S1	0.3565	0.5145	0.1267	0.3256	0.1363	0.3579	0.4292	**0.9360**	0.3753	0.2892
E2.0S2	0.3724	0.5098	0.1350	0.3188	0.1421	0.3100	0.4416	**0.9369**	0.3970	0.4087
E2.0U1	0.6562	0.2789	0.4127	0.1706	0.4191	0.3415	0.4749	0.3455	0.4918	**0.8681**
E2.0U2	0.5698	0.2339	0.4279	0.1176	0.3816	0.2774	0.3864	0.3006	0.4587	**0.9081**
E2.0U3	0.5263	0.1911	0.3590	0.0630	0.3305	0.2026	0.4197	0.2418	0.4811	**0.9030**
E2.0U4	0.5398	0.2700	0.3932	0.1356	0.3469	0.2487	0.4418	0.2965	0.4766	**0.8938**
E2.0U5	0.4415	0.2688	0.3360	0.1663	0.2997	0.2774	0.4378	0.3070	0.5424	**0.8703**
E2.0U6	0.4524	0.3758	0.2821	0.2769	0.3140	0.3252	0.4562	0.4710	0.5569	**0.8249**
OIB1	0.1059	0.5439	0.0431	**0.9422**	0.1783	0.4542	0.1499	0.2851	0.1478	0.0866
OIB2	0.2024	0.5825	0.1088	**0.9533**	0.1921	0.5099	0.2835	0.3632	0.2126	0.2403
PIB1	0.3711	0.5178	0.3729	0.4286	0.4313	0.7388	0.3618	0.2184	0.3951	0.3312
PIB2	0.3133	0.4920	0.2117	0.3861	0.2274	**0.8162**	0.2744	0.2997	0.2095	0.2324
PIB3	0.2094	0.3282	0.1237	0.3084	0.2279	**0.7173**	0.2838	0.3118	0.2144	0.2397
PIB4	0.1538	0.3745	0.0969	0.4037	0.2214	**0.7664**	0.2521	0.2767	0.0926	0.1361
KME1	0.2635	**0.8347**	0.3218	0.4475	0.3050	0.4607	0.1857	0.4378	0.1697	0.3177
KME2	0.2795	**0.9149**	0.2764	0.5912	0.2211	0.5778	0.2584	0.4683	0.1877	0.3127
KME3	0.1891	**0.8991**	0.1991	0.5334	0.1646	0.4870	0.2031	0.4849	0.1795	0.2087
KME4	0.2644	**0.8550**	0.2759	0.5048	0.2591	0.4977	0.2436	0.5238	0.2973	0.2397
LTO1	0.4377	0.3224	**0.8856**	0.1429	0.6719	0.3510	0.3377	0.1135	0.3437	0.3629
LTO2	0.2787	0.2332	**0.8945**	0.0774	0.5714	0.2250	0.2204	0.1081	0.2419	0.2935
LTO3	0.3766	0.2563	**0.8019**	0.0105	0.4841	0.1382	0.1551	0.1365	0.1809	0.3849
LTO4	0.3428	0.1454	**0.7132**	-0.0067	0.4565	0.2049	0.2835	0.1064	0.2768	0.3742
PD1	0.3816	0.1890	0.5813	0.1552	**0.8500**	0.3248	0.2227	0.0323	0.3877	0.3293
PD2	0.3847	0.2489	0.5252	0.2024	**0.8607**	0.3493	0.2450	0.1564	0.2621	0.2890
PD3	0.4332	0.1777	0.6077	0.0584	**0.8596**	0.3216	0.3121	0.1522	0.3904	0.3704
PD4	0.3423	0.2969	0.5437	0.2503	**0.7754**	0.2749	0.1414	0.1615	0.3130	0.3595

Second, a test of discriminant validity as per Fornell and Larcker (1981) was conducted to ensure that the constructs of Power Distance, Long-Term Orientation, Knowledge Management Environment, Personal Information Behavior, Organizational Information Behavior, E2.0 Technology Sophistication, E2.0 Media Richness, E2.0 Perceived Usefulness, E2.0 Perceived Ease of Use, and E2.0 Use Intention were all distinct. A visual inspection of Table 3 shows that the square root of the average variance shared by items within a construct (shown in bold on the diagonal) exceeds the inter-construct correlations that appear below and beside them. All model constructs satisfied these criteria.

Table 3: Average Variance Extracted and Inter-Construct Correlations

Constructs	EOU	KME	LTO	OIB	PD	PIB	Rich	Soph	UI	Usefl
Enterprise 2.0 Ease of Use	**0.903**									
Knowledge Management Environment	0.285	**0.876**								
Long Term Orientation	0.439	0.305	**0.827**							
Organizational Information Behavior	0.165	0.595	0.082	**0.948**						
Power Distance	0.461	0.269	0.673	0.196	**0.837**					
Personal Information Behavior	0.361	0.579	0.287	0.510	0.381	**0.761**				
Enterprise 2.0 Richness	0.629	0.256	0.300	0.232	0.279	0.392	**0.840**			
Enterprise 2.0 Sophistication	0.389	0.547	0.140	0.344	0.149	0.357	0.465	**0.936**		
Enterprise 2.0 Use Intention	0.569	0.239	0.316	0.192	0.402	0.318	0.627	0.412	**0.894**	
Enterprise 2.0 Usefulness	0.609	0.307	0.421	0.177	0.399	0.319	0.498	0.373	0.570	**0.720**

Third, various tests of convergent validity were performed through an assessment of various quality indices as shown in Table 4. As shown, the AVE value for each latent variable is higher than 0.5, indicating that at least 50% of the variance in each block of indicators can be attributed to the pertinent latent variables (Fornell and Larcker 1981; Chin, 1998). Moreover, the values of Cronbach's alpha exceeding 0.70 shows the internal reliability consistency of construct at an individual level (Gefen et al., 2000). Furthermore, composite reliability values for each construct are higher than 0.70 which is the recommended cut-off to validate the internal reliability consistency relative to all other constructs in the model (Fornell and Larcker, 1981).

Table 4: Convergent Validity Assessment of the Measurement Model

Latent Variables	Convergent Validity Indicators		
	AVE (> 0.50)	Composite Reliability (> 0.70)	Cronbach's Alpha (> 0.70)
Enterprise 2.0 Ease of Use	0.816	0.930	0.887
Knowledge Management Environment	0.768	0.930	0.899
Long Term Orientation	0.684	0.896	0.848
Organizational Information Behavior	0.898	0.946	0.887
Power Distance	0.701	0.903	0.857
Personal Information Behavior	0.579	0.846	0.762
Enterprise 2.0 Richness	0.706	0.906	0.861
Enterprise 2.0 Sophistication	0.877	0.935	0.860
Enterprise 2.0 Use Intention	0.800	0.941	0.916
Enterprise 2.0 Usefulness	0.772	0.953	0.941

In order to estimate the structural model, this study used a bootstrapping procedure to derive t-statistics for the structural paths and assess the significance of the path beta coefficients in the structural model. Specifically, bootstrapping with 1000 replications was performed to provide a more conservative test of parameter significance (Chin, 2001). The structural model and the p-values are presented in Figure 3 with results from the two rounds of validation depicted along each path in order of Saudi Arabia and North America (values for the latter also shown as italicized). The main conclusion drawn from the reported structural model analysis as depicted in Figure 3 is that most of the proposed paths are supported with high degrees of confidence.

As predicted (P1), Power Distance had a significant positive effect on Personal Information Behavior in both models. However, no significant relationship emerged between Power Distance and Knowledge Management Environment (P2). P3, concerning the hypothesized effect of Long-

Term Orientation, was only supported for the North American dataset, but did not reach significance with the Saudi Arabia sample. E2.0 Technology Sophistication had significant positive effects on Knowledge Management Environment in both samples, thus supporting P4. Also, as predicted in P5, E2.0 Technology Sophistication had significant direct impact on E2.0 Media Richness in both samples. P6, concerning the hypothesized effect of E2.0 Technology Sophistication on Usefulness, was supported for the Saudi Arabia dataset, but not for the North American sample. P7 and P8 pertaining to the effect of E2.0 Media Richness on Perceived Ease of Use and E2.0 Use Intention for KM was supported for both samples. Similarly, the effect of Perceived Ease of Use on Perceived Usefulness (P9) was supported in both datasets. Proposition P10 which hypothesized a relationship between Knowledge Management Environment and Personal Information Behavior was only supported in the Saudi Arabia dataset. Contrary to expectations, Knowledge Management Environment did not have a direct positive effect on Enterprise 2.0 Use

Intention for KM (P12) in either model. However, a test of mediation showed that for both datasets, Organizational Information Behavior fully mediates this effect. As anticipated, the effect of Perceived Usefulness of E2.0 technologies on E2.0 Use Intention (P13) was supported in both datasets. P14, concerning the hypothesized effect of Personal Information Behavior on E2.0 Use Intention for KM was supported for the North American dataset, but not for the

Saudi Arabia sample. The proposition concerning the direct positive effect of Organization Information Behavior on Personal Information Behavior (P15) was supported for the North American sample but not for the Saudi Arabia sample. Finally, while Organizational Information Behavior had a significant effect on E2.0 Use Intention for KM (P16) in both models, for the North American sample, a negative path coefficient emerged.

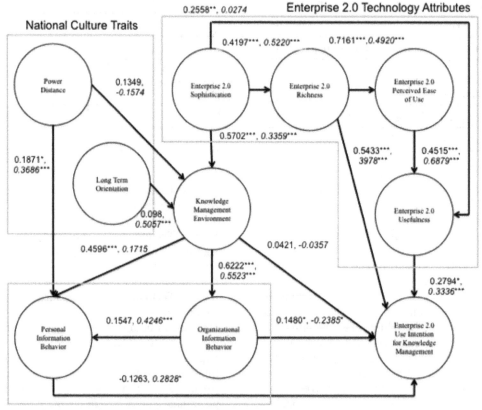

***p<0.001, **p<0.01, *p<0.05, --- not significant at 0.05 level

Figure 3: Theoretical Model Path Coefficients

In determining the efficacy of the model in terms of predictability, an evaluation of the coefficients of determination (R^2) suggested that the model performed well for most downstream endogenous variables. The

coefficients of determination (R^2) explain the proportion of a construct's variance that can be predicted by antecedent constructs in the model. While there is no specific cut-off value for measuring R^2, generally higher values are considered favorable (Gefen et al., 2000), and some researchers suggest that values higher than 0.10 can be deciphered to indicate the usefulness of an endogenous variable within the model (Falk & Miller, 1992). Table 5 below provides the values of the coefficient of determination for all inner model constructs. As seen, most variables compellingly exceed the minimum level of 0.10, with the exception of E2.0 Media

Richness for which the value is only slightly above 0.10. This should not be a concern since this construct only has one incoming path from other constructs that act as their antecedents. In terms of the ultimate criterion variable in the model, i.e. E2.0 Use Intention for KM, a significant portion of its variance (around 52%) can be explained by the model for both datasets.

Table 5: Coefficients of Determination (R^2) for Model Constructs

Model Constructs	Coefficients of Determination (R^2)	
	Saudi Arabia Dataset	North America Dataset
Enterprise 2.0 Ease of Use	0.513	0.242
Knowledge Management Environment	0.400	0.360
Long Term Orientation	na (exogenous)	na (exogenous)
Organizational Information Behavior	0.387	0.305
Power Distance	na (exogenous)	na (exogenous)
Personal Information Behavior	0.421	0.461
Enterprise 2.0 Richness	0.176	0.272
Enterprise 2.0 Sophistication	na (exogenous)	na (exogenous)
Enterprise 2.0 Use Intention for KM	0.529	0.511
Enterprise 2.0 Usefulness	0.364	0.486

For assessing the goodness of fit for the structural model, we used the global criterion of goodness-of-fit (0 <= GoF <=1) to evaluate the model fit (Tenenhaus et al., 2005). This test is defined as the geometric mean of the average communality (AVE) and the average of the R^2 values (for endogenous constructs) (Tenenhaus et al., 2005; Wetzels et al., 2009). For the two rounds of validation, the resulting GoF values of 0.563 and 0.522 both exceeded the cut-off value for large effect sizes (R^2) of 0.35. Hence, it can be inferred that the structural model performed well in both rounds of validation.

Discussion

The findings outlined in the previous section confirm the general premise that national cultural traits, organizational knowledge management context, and enterprise 2.0

technology attributes play an important role in determining the knowledge management practices of organizations, and consequently, the use of enterprise 2.0 technology for knowledge management. As such, we would recommend that knowledge management researchers further explore and investigate these sociological and technological constructs related to the use of E2.0 technology for KM to gain a better understanding of the differences among adoption and acceptance of these technologies. Likewise, practitioners should pay more attention to ensuring that relevant E2.0 tools are available according to the needs and preferences of the organization and its employees. Furthermore, they should ensure the presence of a conducive corporate-wide knowledge management environment comprising effective practices, policies and processes for KM initiatives.

In terms of similarities and differences among national culture traits, our results support the proposition that greater power distance leads people to take it upon themselves to communicate internally and share information, and it does not support a positive knowledge management environment in the enterprise. This was validated in both datasets implying that a hierarchical centralized structure does not support a collaborative knowledge culture. For long-term orientation, our results were inconclusive in that its link to knowledge management environment was not supported in the Saudi Arabian context. This may be explained through complementary findings in our survey that suggested that for the most part, strategies and practices related to knowledge management are still in their infancy in that country. Hence, responses to our survey did not validate a significant positive or negative link between long-term orientation and knowledge management environment.

Generally, our results indicated that KM oriented strategies and the use of E2.0 for KM specific purposes are in their early stages in Saudi Arabia, while in North America, processes, practices, policies and technologies related to knowledge management are more mature. This finding has multiple implications and helps explain some of the results from our structural model.

Firstly, personal information behaviors in Saudi Arabia are not impacted by organizational information behaviors since the information and knowledge sharing practices at the corporate level are still not well established in these firms. This outcome is also supported through responses to our technographic questions which indicated that most individuals in Saudi Arabia use social media tools for personal internal communication purposes rather than collaboration and knowledge sharing purposes.

Secondly, in terms of technology use, it can be seen that in the context of North America, both Organizational and Personal Information Behaviors affect intention to use E2.0 tools for KM purposes, while in the case of Saudi Arabia, only the former was a significant determinant. This shows that personal information sharing and communication behaviors are not sufficient to sustain the use of E2.0 technologies for KM, and organizations really need to foster a corporate-wide culture of information sharing and knowledge transfer in order to stimulate and promote the use of enterprise 2.0 technologies for knowledge management purposes. Based on the path coefficient values in the structural model, it can also be argued that Organizational Information Behavior may play a more important role in the development period of KM strategy formulation and the early days of E2.0 technology use for KM. This is evident in the positive significant effect of organizational information behavior on enterprise 2.0 use for knowledge management in the context of Saudi Arabia. Once a favorable knowledge management culture is present, people may take the initiative on their own to improve the use of enterprise 2.0 technologies for knowledge management purposes – as observed in the significant positive impact of personal information behavior on E2.0 use for KM in the Canada and US dataset. Flexibility in use of corporate social software has been regarded as a primary driver for using such tools for knowledge management (Richter et al., 2013).

Thirdly, with respect to the variety of features and functions of enterprise 2.0 tools, our model showed a significant positive relationship between E2.0 sophistication and perceived usefulness for the Saudi Arabia dataset but not the North American dataset. Once again, this may be attributed to the embryonic nature of KM strategies and technologies in the Saudi Arabia context at the current time. Organizations in such a setting may still be experimenting with a variety of social media tools to understand and cater to employee needs and requirements of their knowledge intensive processes, and hence the availability of a breadth of social media channels and platforms translates into higher perceptions of usefulness. In contrast, in the North American context, the E2.0 technology base may be more evolved and stable, and variety does not necessarily translate into higher levels of perceived usefulness.

As far as the enabling role of E2.0 technologies for KM is concerned, our model clearly showed a significant positive impact of E2.0 technology sophistication and knowledge management environment in both datasets. This highlights the value and utility of these technologies in promoting a knowledge sharing corporate culture by facilitating organized and accessible information, and enabling the sharing of best practices across the organization. Furthermore, it can be argued that the relationship among E2.0 technologies for KM and the organizational KM environment is reciprocal and mutually reinforcing. A test of mediation was conducted to evaluate the relationship between the organizational KM environment and the intention to use E2.0 tools for KM. Organizational Information Behavior was shown to fully mediate the effect of organizational KM environment on the intention to use E2.0 tools for KM. This result is noteworthy since it was supported for both datasets suggesting that the adoption and acceptance of KM technologies is a gradual process that requires the presence of a favorable knowledge management culture in the organization as well as visible and recognizable information sharing and knowledge collaboration practices across the enterprise.

Our results lend strong support to the knowledge management context model proposed by Detlor et al. (2006) where the researchers purported links between knowledge management environment, personal information behavior, and organizational information behavior. As noted above, these links were validated in either one or both of our datasets.

Furthermore, through our work, we have attempted to answer the call for additional research by the authors of the original model. Our model extends their work by offering additional sociological and technological constructs that can help explain the variance in organizational and personal information behaviors. As shown in Table 5, our model is able to account for the variance of organizational information behavior in the range of 0.31 (North America) to 0.40 (Saudi Arabia), and in the range of 0.42 (Saudi Arabia) to 0.46 (North America) for personal information behavior. This is in contrast to the variance for these variables reported in the range of 0.17 to 0.19 in the original model by Detlor et al. (2006).

In terms of limitations, this study is constrained by the investigation of knowledge management practices and enterprise 2.0 technology use using convenience and self-selection sampling techniques. This may limit the generalizability of the study's results. However, to some extent, the validation of the theoretical model in multiple national and industrial contexts mitigates this limitation.

Future studies wishing to explore further the effects of national and organizational culture, and technology characteristics on knowledge management practices and technology use may wish to adopt this study's research model as a theoretical basis. The model confirms and extends Detlor et al.'s (2006) model and highlights the importance of knowledge management culture and information behaviors at organizational and individual levels on the overall effectiveness of KM practices and the use of enterprise 2.0 technologies for KM purposes. Researchers also may wish to carry out the investigation in additional regional and industrial contexts for purposes of comparison. Finally, we believe it would be very fruitful to conduct a focused investigation using case-studies of organizations that may be in different stages of use of enterprise 2.0 tools for knowledge management.

Conclusion

This paper presented research results of an empirical investigation of sociological and technological factors affecting knowledge management practices and the use of enterprise 2.0 technologies for knowledge management in organizations. Results indicate that national culture traits and corporate KM culture play an important role in influencing personal and organizational information behavior, as well as the use of enterprise 2.0 technologies for KM. Additionally, the availability of a variety of enterprise 2.0 tools is critical in the early stages of KM strategy development as these tools help foster a positive knowledge management environment in the organization.

The extension of Detlor et al.'s model with national culture traits and technology attributes has important practical and theoretical value, but additional research is needed. As organizations continue to experiment with social media channels and platforms for knowledge sharing and collaboration, they also need to continue their efforts to improve the culture, processes and practices around information sharing, and the transfer of embedded and encoded knowledge across the enterprise.

References

1. Alavi, M., & Leidner, D. E. (2001). Review: Knowledge management and knowledge management systems: Conceptual foundations and research issues. MIS Quarterly, 25(1), pp. 107-136.

2. Alhamoudi, Salwa. (2010). Strategic KM system in public sector in Saudi Arabia: an adaptation of the Balanced Scorecard. PhD thesis, University of Portsmouth.

3. Anderson, J. C., & Gerbing, D. W. (1988). Structural Equation Modeling in practice: A review and recommended two-step approach. Psychological Bulletin, 103(3), 411-423.

4. Andrews, D., Preece, J., and Turoff, M. "A Conceptual Framework for Demographic Groups Resistant to Online Community Interaction," 34th Annual Hawaii International Conference on System Sciences, Maui, Hawaii, 2001.

5. Ardichvili, A., Cardozo, R. and Ray, S. (2003). A Theory of Entrepreneurial Opportunity Identification and Development, Journal of Business Venturing, 18/1, pp. 105-123.

6. Boyd, Michael (2012). Hofstede's Cultural Attitudes Research - Cultural Dimensions, Retrieved March 20th, 2012 http://www.boydassociates.net/Stonehill/Global/hofstede-plus.pdf

7. Chin, W. W. (1998). Issues and opinion on structural equation modeling. Management Information Systems Quarterly, (22:1), pp. 7-16.

8. Chin, W.W. "PLS-graph user's guide version 3.0," Soft Modeling Inc, 2001.

9. Daft, R. L., & Lengel, R. H. (1986). Organizational information requirements, media richness and structural design. Management Science, 32(5), pp. 554-71.

10. Davenport, Thomas H., Prusak, Laurence (2000). Working Knowledge: How Organizations Manage What they Know. Harvard Business School Press, pp. 240. ISBN 1-57851-301-4.

11. Davis, F. D. (1989). Perceived usefulness, perceived ease of use, and user acceptance of information technology. MIS Quarterly, 13(3), pp. 319-340.

12. Davis, F. D., Bagozzi, R. P., & Warshaw, P. R. (1989). User acceptance of computer technology: A comparison of two theoretical models. Management Science, 35, pp. 982–1003.

13. DeLone, W. H. and McLean, E. R. (1992). Information Systems Success: The Quest for the Dependent Variable. Information Systems Research, Vol. 3, No. 1, pp. 60-95.

14. DeLone, W. H. and McLean, E. R. (2003). The DeLone and McLean Model of Information Systems Success: A Ten-Year Update. Journal of Management Information Systems, Vol. 19, No. 4, pp. 9-30.

15. Detlor, B., Ruhi, U., Turel, O., Bergeron, P., Choo, C. W., Heaton, L., & Paquette, S. (2006). The effect of KM context on KM practices: An empirical investigation. The Electronic Journal of Knowledge Management, 4(2), pp. 117-28. Available at www.ejkm.com

16. Falk, R.F., and Miller, N.B. (1992). A primer for soft modeling University of Akron Press Akron, OH.

17. Fornell, C., and Larcker, D.F. "Structural equation models with unobservable variables and measurement error: algebra and statistics," Journal of Marketing Research) 1981, pp 382-388.

18. Gefen, D., Straub, D. W., & Boudreau, M.-C. (2000). Structural Equation Modeling and Regression: Guidelines for research practice. Communications of the Association of Information Systems, 4(7), 1-77.

19. Ghobakhloo, M., Benitez-Amado, J., & Arias-Aranda, D. (2011a). Reasons for information technology adoption and sophistication within manufacturing SMEs. Paper presented at the POMS 22nd Annual Conference: Operations management: The enabling link. Reno, USA, April 29 to May 2, 2011.

20. Hazlett, S.A., McAdam, R. and Gallagher, S. (2005). "Theory building in knowledge management: in search of paradigms." Journal of Management Inquiry, Vol. 14 No. 1, pp. 31-42.

21. He, W. and Wei, K. K. (2009). What drives continued knowledge sharing? An investigation of knowledge contribution and -seeking beliefs. Decision Support Systems, Vol. 46, No. 4, pp. 826-838.

22. Hofstede, G. (1980). Culture's Consequences: International differences in work related values. Beverly Hill, CA, Sage.

23. Hofstede, G.H. (1997). Cultures and Organizations: Software of the mind. London. McGraw-Hill.

24. Holden, N.J. (2002) Cross-cultural Management: A Knowledge Management Perspective. London: Financial Times/ Prentice Hall.

25. Holsapple, C. W. & Joshi, K. D. (2000). An investigation of factors that influence the management of knowledge in organizations. Journal of Strategic Information Systems, 9(2/3), pp. 235-261.

26. Kane, J., Robinson-Combre, J., and Berge, Z. L. (2009). Tapping into social networking: Collaborating enhances KM and e-learning. Journal of Information and Knowledge Management Systems. Vol. 40, No. 1, pp. 62-70.

27. Kuhlen, R. (2003), "Change of paradigm in knowledge management – framework for the collaborative production and exchange of knowledge." Paper presented at the 69th IFLA General Conference and Council, 30 August 2003, Berlin.

28. Kim, S. and Lee, H. (2006). "The impact of organizational context and information technology on employee knowledge-sharing capacities." Public Administration Review, Vol. 66 No. 3, pp. 370-85

29. Malone, T. W. (2003). Organizing business knowledge: The MIT process handbook. Cambridge, MA: MIT Press.

30. Marquardt, M. and Kearsley, G. (1999). Technology-based learning: Maximizing human performance and corporate success. Boca Raton: CRC Press LLC.

31. McAfee, A. P. (2006). Enterprise 2.0: The dawn of emergent collaboration. Engineering Management Review, IEEE, 34(3), pp. 38-38.

32. Nonaka, I. and Takeuchi, H. (1995). The Knowledge-Creating Company: How Japanese Companies Create the Dynamics of innovation. New York: Oxford University Press.

33.O'Reilly, T. (2005). What Is Web 2.0. Design Patterns and Business Models for the Next Generation of Software. Retrieved November 16th, 2011. http://oreilly.com/web2/archive/what-is-web-20.html.

34.Pauleen, D. (2007). Cross-cultural perspectives on knowledge management. Westport, Conn., Libraries Unlimited.

35.Pawlowski, J. M., & Bick, M. (2012). The global knowledge management framework: Towards a theory for knowledge management in globally distributed settings. Electronic journal of knowledge management, 10(1), pp. 92-108.

36.Podsakoff, P.M., MacKenzie, S.B., Lee, J.Y., and Podsakoff, N.P. "Common method biases in behavioral research: A critical review of the literature and recommended remedies," Journal of Applied Psychology (88:5) 2003, pp 879-903.

37.Podsakoff, P.M., and Organ, D.W. "Self-reports in organizational research: Problems and prospects," Journal of Management (12:4) 1986, p 531.

38.Richter, A., Stocker, A., Müller, S., & Avram, G. (2013). Knowledge management goals revisited: A cross-sectional analysis of social software adoption in corporate environments. *VINE*, *43*(2), 132-148.

39.Riege, A. (2005). Three-dozen knowledge-sharing barriers managers must consider. Journal of KM, Vol. 9, pp. 18-35.

40.Ringle, Christian M., Wende, Sven, & Becker, Jan-Michael. (2013). Smartpls 3. Hamburg: SmartPLS. Retrieved from http://www.smartpls.com

41.Ruhi, U. and Choi, D. (2013). Enterprise Mashups for Knowledge Management. Proceeding of 1st International Conference on Information and Communication Technology Trends (ICICTT).

42.Saffady, W. (1998). Knowledge management: A manager's briefing. Prairie Village, KS: ARMA International.

43.Stankosky, M. (2005). Creating the discipline of knowledge management: the latest in university research.Amsterdam. Boston, Elsevier Butterworth-Heinemann.

44.Tenenhaus, M., Vinzi, V.E., Chatelin, Y.M., and Lauro, C. (2005). "PLS path modeling." Computational Statistics and Data Analysis (48:1), pp. 159-205.

45.Thomas, R. D., Lu, I. R. R., & Cedzynski, M. (2005). Partial Least Squares: A critical review and a potential alternative. Proceedings of the Administrative Sciences Association of Canada (ASAC) Conference, Toronto, Ontario, Canada.

46.Velev, Dimiter and Zlateva, Plamena. (2012). Current State of Enterprise 2.0 Knowledge Management. International Journal of Trade, Economics and Finance, Vol. 3, No. 4.

47.Voelpel, S. C. & Han, Z. (2005). "Managing knowledge sharing in China: the case of Siemens ShareNet." Journal of Knowledge Management, 9, pp. 51-63.

48.Wetzels, M., Odekern-Schroder, G., and Oppen, C.v. (2009) "Using PLS Path Modeling for Assessing Hierarchial Construct Models: Guidelines and Empirical Illustration." MIS Quarterly (33:1), pp. 175-199.

49.Zheng, Y., Li, L. and Zheng, F. (2010). Social Media for Knowledge Management. Knowledge Creation Diffusion Utilization, pp. 9-12.

Appendix A: Measurement Items for Theoretical Constructs

Theoretical Construct	Measurement Items
Power Distance	(7-point Likert Scale from Not Important to Very Important) - Having a good working relationship with your direct superior - Being consulted by your direct superior in his/her decisions - Being consulted by your boss in decisions involving your work - Relations between a superior and subordinates are open & friendly - Authority to match position or role in the hierarchy - Authority to match competencies
Long-Term Orientation	(7-point Likert Scale from Not Important to Very Important) - Working toward future goals - Working for future life - Saving money for future - Persistence (perseverance) - Steadiness and stability - Respect for tradition
Perceived Ease of Use	(7-point Likert Scale from Strongly Disagree to Strongly Agree) - I find it easy to get Enterprise 2.0 tools to do what I want them to do. - It is easy for me to become skillful at using Enterprise 2.0 tools. - I find Enterprise 2.0 tools easy to use.
Perceived Usefulness	(7-point Likert Scale from Strongly Disagree to Strongly Agree) Using Enterprise 2.0 technologies allows me to: - Accomplish tasks more quickly - Improves my job performance - Increases my productivity - Enhances my effectiveness on the job - Makes it easier to do my job - Overall, I find Enterprise 2.0 useful in my job
Media Richness	(7-point Likert Scale from Strongly Disagree to Strongly Agree) Enterprise 2.0 technologies allow me to: - Tailor interactions according to my personal requirements and preferences - Communicate a variety of different cues (such as emotional tone, attitude, or formality) during communication - Use varied and rich language during communication - Convey multiple types of information
Technology Sophistication	(7-point Likert Scale from Strongly Disagree to Strongly Agree) - I have access to a wide variety of Enterprise 2.0 tools at my workplace. - I have all the essential Enterprise 2.0 tools at my workplace.
Knowledge Management Environment	(7-point Likert Scale from Strongly Disagree to Strongly Agree) - My organization has a culture intended to promote knowledge and information sharing. - Knowledge and information in my organization is available and organized to make it easy to find what I need. - Information about good work practices, lessons learned, and knowledgeable persons is easy to find in my organization. - My organization makes use of information technology to facilitate knowledge and information sharing.

Theoretical Construct	Measurement Items
Organizational Information Behavior	(7-point Likert Scale from Strongly Disagree to Strongly Agree) - The people I work with regularly share information on errors or failures openly. - The people I work with regularly use information on failures or errors to address problems constructively. - Among the people I work with regularly, it is normal for individuals to keep information to themselves.
Personal Information Behavior	(7-point Likert Scale from Strongly Disagree to Strongly Agree) - I often exchange information with the people with whom I work regularly. - I often exchange information with people outside of my regular work unit but within my organization. - I often exchange information with citizens, customers, or clients outside my organization. - I often exchange information with partner organizations.
E2.0 Technology Use Intention	(7-point Likert Scale from Strongly Disagree to Strongly Agree) - If I have access to Enterprise 2.0 tools, I predict that I would use it - I intend to use Enterprise 2.0 tools as often as needed - I intend to continue using Enterprise 2.0 tools for social interactions - I intend to continue using Enterprise 2.0 tools for information exchange
E2.0 Technology Actual Use	(7-point Likert Scale from Use Rarely to Use Frequently) Social Networking Sites (e.g. Facebook) Blogs Wikis Discussion Forums Content Feeds (e.g. RSS) Collaborative Document Editing Collaborative Content Tagging Social Bookmarking Mashups Instant Messaging Microblogs (e.g. Twitter) Polls & Voting Tools Widgets File Sharing (Repositories) Podcasts Video Sharing Virtual Worlds Online games Social News

12

Knowledge Management and Management Accounting Decisions- Experimental Study

Afaf Mubarak

Faculty of Business Administration[1], Al Hosn university- Abu Dhabi- United Arab Emirates

Abstract

In this study, the author explores how individuals in the UAE who involve in making management accounting decisions (e.g. planning, investment, allocating resources,...etc) would acquire, distribute and use knowledge. The study replicates the experiment of Edward et.al. (2006) of CIMA project about how knowledge is managed within management accounting decisions. A total of 26 participants from different departments and different industries (hotel, insurance, retail, education, travel and food) contributed in answering questions on cases present routine and strategic decisions. The results indicate that participants do not differentiate between information and knowledge and all organizations have effective methods for collecting and maintaining knowledge in databases, files and other documents and different information are kept and used for the different decisions. In general, contributors in the study tend to reply more on formal, written information both financial and non-financial. However, for strategic decisions the non-financial information from different sources tend to gain more importance. Local respondents and those work in environments which encourage risk-taking tend to depend on informal and oral information more than non-local participants and those in middle management. Knowledge for routine decisions are processed via computerized programs but participants could not explain how they process knowledge in their minds and how they utilize in the different situations. The study revealed that there are differences in knowledge and its management according to industry, managerial level and type of decision and there were reflections of the UAE's business environment with its multi-culture and having specific strategies for example to encourage creativity as in tourism industry or to be tightly regulated as insurance industry.

Key words: knowledge management- accounting- experiment – UAE.

---

[1] The author would like to thank participants of IBIMA 18th conference in Turkey –May 2012 and an anonymous reviewer for constructive comments.

Introduction

Since the "relevance lost" of management accounting which has been framed by Robert Kaplan and Thomas Johnson (1987), and the discipline has been addressed from different perspectives and in different ways. The intention is to improve its grounds and to provide better understanding and analysis for the different phenomena it addresses. One of the old and ever visited approaches is to investigate management accounting as an information system and within that approach, to analyze it under specific lenses. One of those latter stands is to examine the decision making process in management accounting with a "Knowledge Management" lenses. While there is no consensus on what knowledge or knowledge management (KM) is, the author referred to a study of Firestone (2001) which surveyed and analyzed a good deal of definitions for both. However, it is not among the targets of this study to evaluate the different definitions of knowledge nor to compare it with data and information or to do the same with KM. Instead, the author will adopt a functional definition of knowledge management that fits with the type of the current study. The aim of this study is to explore how KM is used in making decisions and to do so, the functional definition will be consistent with the decision making process generally. The definition adopted for Knowledge Management in this study is that it consists of activities focused on the organization gaining knowledge from its own experience and from the experience of others, and on the judicious application of that knowledge to fulfill the mission of the organization. Knowledge management (KM) is the name of a concept in which an enterprise consciously and comprehensively gathers, organizes, shares, and analyzes its knowledge in terms of resources, documents, and people skills. The definition of "Knowledge" adopted is that; while knowledge is made up of data and information, can be thought of as much greater understanding of a situation, relationships, causal phenomena, and the

theories and rules (both explicit and implicit) that underlie a given domain or problem."

Knowledge management has been receiving a continuous and growing attention on academic and practical levels. On academic level, it is at the top of themes of discussion in most conference of business and information systems. In addition, many journals and publications are allocated to examine KM from different perspectives e.g. Taylor 2006, Benbya 2008, Maier 2007, Bhatt 2001. Examples of discussing accounting and KM, are O'Leary (2002) who wrote a chapter about knowledge management in accounting and professional services, Brikett introduced as early as 1995, a case study of knowledge management as a framework for studying strategic decisions in accounting and Sori (2009) examined the use of accounting information systems by Malaysian companies and its contribution to the knowledge management and strategic role in the organization. Bhimani and Roberts (2004) raised a question about intelligibility in management accounting and knowledge management and Sabina-Cristina (2007) paid an attempt to model accounting decisions relates to fixed assets within knowledge management, study applied in Romania and Booth et.el. (2009) examine two factors that impact managers' willingness to share private information (which is one step of knowledge management activities) during the project review stage of capital budgeting. On practical levels, companies seek to provide training programs to their employees in how to manage their knowledge effectively and also they request consultancy advices about how to build up effective KM systems and Business environment in UAE is not an exception of that. The following section highlights why I chose to study this topic in this country.

Why KM in the UAE

UAE is a case of emerging economy that although rich from oil exporting, tries to diversify economic activities. The country tries to move away from depending on a sole source of income which is oil, by expansion off-shore in some activities such as telecommunications, construction, retail and leisure industries. In doing so, businesses (private and public drive their steps by adopting most up-to-date international business practices and one of that is to build up a supportive knowledge management systems. Just examples of concern are; the conference of "KM Middle East, 13-14 March 2012" in the capital city of Abu Dhabi and "UAE e-Government Employee Knowledge Management Development Conference" Dubai 5-7July 2011 and "International Human Resources Conference- Knowledge Management: Challenges and Applications in the GCC" 19-20 January 2011 Dubai-UAE. The second conference reflects an intention to imbed KM into the development of e-government in UAE and the latter one apparently focused on KM practices and improvements within the region of Gulf Co-operative Council (GCC) to which UAE belongs.

KM gains a special importance in UAE for particular reasons as the author noticed. First, there are giant businesses with billions of dollars resources (e.g. Etisalat telecommunication had total assets of 20b dollar and size of operations about 11.8b dollar in 2011 and operations in 17 markets; Emmar for real estate has total assets of 17.3b dollar in 2010 and operates in many countries in MENA, Europe and Asia). Re-export valued more than 60b dollar according to the National Bureau of Statistics in UAE in 2010. This suggests that there are huge body of knowledge have to be run effectively and efficiently and it is expected that a reliable management of knowledge is a key factor of success here. Secondly, UAE is known to have business relations with many countries around the world (as indicated in the case of Etisalat telecommunications which operates in many countries including Egypt, Saudi Arabia, Sudan, Tanzania and others, Emmar operates in Egypt, Morocco, UK, Asian countries and the global holding company Dubai World works in Transport & Logistics, Drydocks & Maritime, Urban

Development, Investment & Financial Services, its portfolio extends from the Americas to Asia including India, China and the Middle East). Doing work in and with foreign countries –in addition to the government's genuine initiative to build a modern country- urged UAE to adopt the most modern and up-to-date business practices in every aspect; e.g. corporate governance, disclosure practices and KM. The third reason in my view is the fact that most of the work force in UAE is expatriates who came from different countries for work. According to Bureau of National Statistics, the foreign employment makes more than 60% of the work force from many countries of which 80% are Asian, 15% Arab and the remaining belong to different countries. Having a business environment that is multi-culture raised the importance of having smooth and effective methods and tools for communicating and distributing knowledge among co-workers to help them work, cooperate and create more smoothly at work.

Knowledge Management and Decisions' Making in Accounting

While there are many dimensions in KM topic to be discussed, the focus in this study is on knowledge management while making decisions in management accounting domain. Accounting plays different roles in organization. In base, it is an information provider to decision makers and in doing so, accounting follows set of steps include collection of its own information, analyze and organize those information and suggest or recommend a solution or alternative. Accounting does involve in many types of decisions for example; planning by preparing budget, evaluating performance, measure costs and revenue, run sensitivity analysis for volume of production, costs and profit, making investment decisions and others. When accounting is doing its functions, it generates, stores, analyze and utilize a body of "knowledge" that is added to the overall knowledge of organization. The purpose of this study is to explore how accounting information are developed and used as part of the body of the knowledge of the organization. It is known that accounting helps is providing information to different users; internal and external to the organization and for different decisions; operational and strategic.

Wilkinson (2000) considers the management team that consists of Finance General Manager, Chief Operating Officer, Managing Director and Board of Directors are among the internal users of the system. On the other hand, the external users consist of government agency (e.g. tax authority), external auditors and creditors. Indeed, wide variety of people within and outside the organization uses accounting information for decision-making. Accounting information contribute to the profession's value added to the organization. In addition different decisions are supposed to have specific needs of information. Courtney (2001) reviews various studies that debate how organizations manage knowledge with the purpose of developing some decision support system. It could be viewed from different perspectives, procedurally knowledge is created in the heads of people, captured and put on paper, in a report or a computer system or a library. Then knowledge is classified or modified so that it can be restored by others. When knowledge is re-used, a new knowledge may be created by putting it into new context, background or into new analysis. This is valid especially in social sciences such as accounting where knowing and knowledge are inseparable from action. Knowledge is viewed as an object and an action and organizational knowledge is viewed as a "collective mind" developed through interpretation, communication and shared meaning.

Sori (2009) argues that both tacit and explicit knowledge are used as shown by the extensive used of accounting information system to assist business decision-making. While there are different jobs for management accounting, and hence different contributions of accounting information and different processes for this information, this

research targets to examine how knowledge is collected, processed, distributed and utilized in different decisions' situations; ranging from routine to strategic decisions and their needs of information and possible outcomes. Chou et.al., (2008) illustrate that accounting is developed to demonstrate an architecture of ontology. the development of accounting ontology presents a framework for building accounting knowledge and may also serve as an excellent learning tool for accountants. According to the writers, the process of constructing the accounting ontology, consists of 5 stages: Collect Accounting Information, Analyze Accounting Items, Accounting Item Taxonomy, Import Accounting Items to build up interrelations and Generate Ontology for Accounting to develop the acting architecture. The researcher states that this presentation by Chou et.al. (2008) introduces accounting as a body of "knowledge" in organizations that is developed and used in a way quite similar to that of the general KM activities highlighted earlier and this proves the consistency of the research phenomenon (Management accounting decision) and the framework of analysis (which is Knowledge management) selected in this study.

This study is a kind of replication to the essence of the CIMA research about 'Knowledge Management and its impact on the Management Accountant" by Prof John S Edwards, Dr Paul M Collier and Dr Duncan Shaw published in 2006. The reasons for this duplication stem from the functions which a replication study could fulfill. Thomas (2003) states that a replication study is designed to perform one of four functions: (1) To assess the results of an earlier investigation in order to confirm or disconfirm the reported outcomes of that investigation, (2) to judge how stable the results have remained with the passing of time and to estimate the causes of any changes that occurred, (3) to alter some aspect of the earlier methodology in order to discover what effect such alteration has on the outcome, (4) to apply the earlier method to a different group of people or different set of events in order to learn whether conclusions derived from the earlier study apply equally well to those different people or events. The author believes that functions 1 and 4 mentioned by Thomas are valid in this study. I hoped to test whether the Western-run study of CIMA would provide the same results if repeated in an Arab environment that is described as still developing, has its own culture, practices and its own features. Secondly, I am deliberately interested to find out impacts of some factors in this study such as business environments' of making accounting decisions on the process of knowledge management. Variables of consideration include type of decision (routine or non-routine), needs from information (financial and non-financial, facts, rumor, ...etc), type of industry and impacts of managerial level, business strategy and culture. In conclusion, the contribution of this research would develop from:

a- Doing a replication of a previous study to examine what/how knowledge is obtained, processes, shared and stored in an environment different from the original study (which is the UAE) and on different participants (middle and senior managers from different industries).

b- Examine if specific factors of concern (industry, managerial level, type of decision, strategy and cultural issues) may affect the activities of managing knowledge. This line of investigation was not covered by the original study.

c- Collecting feedback on the research questions directly from individuals –in the different industries- involved in making decisions would reveal if the process of acquiring, processing, sharing and storing knowledge done by organizations is relevant, adequate or not, to what extent, suggestions from participants regarding any developments.

In this research, the focus is on arguments around knowledge management activities within making accounting decisions. The research focus could be formatted into the following themes:

- What is knowledge required for management accounting decisions? Links to information?

- How this knowledge is collected? From which sources? In which format?

- How this knowledge is processed both in the organization's information system and in users' minds? How shared (formally/informally)? How stored?

- What are the effects of type of decision, type of industry, strategy and culture on knowledge management?

Research Methodology

In order to explore the research themes, the author selected to use a semi-structured experiment. Experiment is suggested by Christensen et.al., (2011) to be used in examination when there is a need to 'observe" specific outcome and test a phenomenon under specific conditions. An experiment allows the investigator to employ standardized procedures to investigate the effects of treatments. Such standardization ensured a high internal validity or the ability to attribute findings and this is this is the target in this research. As explained earlier, the UAE is a multi-culture business environment, expanding and has links overseas. Therefore, I intend to explore how individuals involved in accounting decisions such as planning via budgets, measuring costs, preparing financial information and investment decisions, how they collect, distribute, share and use information relate to those decisions.

While there are different types of experiments (true, repeated-measures designs, quasi-experiments, and time series designs). Internal validity is generally highest with true experiments due to the random assignment of subjects to different treatments. Conducting experiments is facilitated by following a systematic planning and application process. A seven-step model suggested by Christensen et.al., (2011) consists of (1) selecting a topic, (2) identifying the research problem, (3) conducting a literature search, (4) stating research questions or hypotheses, (5) identifying the research design, (6) determining methods, and (7) identifying data analysis approaches. In my study, the participants are allocated randomly and a true experiment is run.

Because the author had no access to the research instruments of the original study, she intends to investigate the knowledge management practices when taking decisions in management accounting in different industries in UAE. Using a semi structured experiment, the author targets to examine how the jobs of knowledge management ranging from knowledge collection, classifying, processing and distribution are applied? Are there differences between doing these jobs for a routine decision and a strategic decision? What type of information relates to each, financial, technical,...etc? How information flow, formally, informally? What are the sources of knowledge, internal, external? What do information relate to: people, business solutions or processes? And is there an impact of managerial level of decisions maker on this debate?

In order to collect data, eight workshops were held with accountants and managers from different departments and at different levels. Participants belonged to organizations that works in different industries and of different types (governmental, public and private) works in both the capital Abu Dhabi and the tourism well-known city and what was the economic capital (before the international financial crisis) Dubai. The

number of organizations and industries they work in are:

-Municipality (1)

-Insurance (2)

-Hotel and Leisure (1)

-Retail (2)

-Travel (1)

-Food and beverage producing (1)

-Education (1)

Participants are deliberately chosen from different nationalities as this element could make a difference in the participation of creation, distribution and utilization of knowledge. As mentioned in the earlier introduction section, the National Bureau of Statistics in UAE announces facts about the diversity of workforce in the country and the domination of expatriates. To take this fact into account, participants were chosen from different nationalities, of the total 26 participants, 6 were Emirates, 8 Western, 7 Arabs and 4 Asian. This distribution is not consistent with that of the overall population of workforce mentioned in the introduction. The reason for the difference relates to managerial level. Participants were targeted to be working in middle and senior levels and those are largely filled up with locals, western and Arabs. Given that decisions in management accounting are undertaken not only by accountants but also personnel from different departmental functions involve so participants in this experiment will be working in different departments.

Each workshop included a group of 3-6 participants from the same company with total of 26 participants in 7 sessions. Each group was allocated two jobs; one is a case of a 'routine" planning budget and another job that is rather "strategic" relates to capital investment. Data were collected on a questionnaire with many open questions where participants were asked to record on a computer their evaluation for the case, record their needs of information if they are involved in this decision in reality, they were also asked to classify their needs of information if financial or non, technical or general, relates to regulations or procedures and processes. The participants were also asked to highlight the source of the information they need, internal or external.

This method of collecting data and to enrich the data, the researcher depended on interviews implemented while participants are answering questions. The researcher distributed questions and recorded notes for answers and in the mean time, kept answering enquiries about what is the meaning of a point or another and what is required clearly here or there. The researcher spent a specific effort to attract the attention of participants to report about how they get the information they need for their decisions, how such information are organized by their companies, how communicated and the degree of utilizing those knowledge and information. In answering all questions, participants were encouraged to record all their thoughts, ideas, notes,...etc on the light of their real experience as much as possible. After getting answers, the researcher tried to map what the points of agreements and points of differences between participants and whether some differences stem from difference in managerial level, type of decision or type of organization. The following section reports the key findings and will include a comparison with the results of the original study of Edward et.al. (2006) as well as with studies which examined the same or similar themes in the literature.

Research Findings

The following dimensions could be identified from the participants' feedback:

I: Features of Knowledge

- Differentiation between information and knowledge

- Type of information used: financial, non-financial, technical, facts, rumors or judgment

- Format of information: formal, informal, raw or processed and written or oral

- Source of information: internal, external

II: How knowledge is managed:

● How acquired: specific forms/format, databases and other stores

● How processed

● How distributed and shared

● How used

I: Features of Knowledge

- Differentiation between Information and Knowledge

Almost all participants seemed not concerned with differentiating between data, information and knowledge. One reason was given by a production supervisor in food manufacturing is that users tend to use "information" that has been used before by others, that result from calculation of expected sales in the coming quarter, "*would this be "information? Or data? Or knowledge to me? I don't care, I need that piece of information to be available for me when I need to build up my production schedule accordingly*". In insurance company, policy processor officer records that he needs all regulations and decrees that govern his work to be clear and available although he is aware that those information resulted from discussions and meetings among senior managers who draw the broad lines for the business. So when he uses the previously processed information is it information or data or knowledge? One participant from municipality states that: *In doing my work, I consciously or subconsciously re-call different pieces of information, data, knowledge which I "carry in my mind" in addition to what I use from regulations and past experience documents, and I apply all that in the situation.* A students' affairs' manager in an educational institute considers "knowledge" is "what *is stored in our mind not only data and information we got, but how we processed, what relationships, evaluations, speculations and predictions we make, this is affected by our personality. Therefore, you could teach two people the same information, one may remember all the details, make links, summarizes and generates new knowledge which he/she uses in a way different from the other person.* This lack of distinction between knowledge and information somehow also appeared in the original study by Edward et.al. (2006) and when a differentiation was made, participants considered information as a commodity while knowledge is part of the intellectual capital of the organization. Participants in the latter study also saw that there is a great deal of information and knowledge exist in organizations but not well-managed or shared even with the use of email or meetings because wrong people are communicating or being run ineffectively.

The author finds the lack of a clear distinction mentioned, is consistent with the definition of knowledge adopted in this study and presented in the introduction section. That is, knowledge, includes raw data, processed information and experiences used in a situation. The previous views mentioned by participants miss a point, that any "information" could be "data" in a further processing operation. Therefore, in the remaining of this section, no distinction will be made between data, information or knowledge.

Type of Information Used

Participants report different types of information they use in their job, according to everyone's field of work and to type of decision undertaken. For example, participants in finance and accounting

departments depend largely and mainly on financial measures relate to costs, revenues and values of resources. However, marketing manager in retail company is concerned not only with the amount of sales but also the distribution of those sales over seasons. He explains: "*We are working in a country where shopping is a genuine habit to the Emirati people and to the expats. The Empirati loves getting new items in festivals, occasions and when going for holidays. Expats' high purchase season is in summer when they go home. They love to take presents to family and friends from their host country. I can't confine sales to those seasons, my target is to promote sales all over the year and I try to provide different techniques of promotion, therefore I depend on related-products, promotion for volume of sales, introducing different prices....etc, so financial and non-financial data are important to me*". In insurance company the officers of processing policies highlight that different databases exist in their system; databases for customers, items insured, cost of insurance, compensation paid under each policy and records for damaged happened during each policy. In hotel, events and banquet manager maintains organized files for customers, events and their requirements as per customer's specification and steps executed in each event, his resources in-house and resources that can be outsources and details of them. For this manager, financial information such as prices and costs are of limited importance (only needed to price the contract). An engineer works as deputy road maintenance of an area of Abu Dhabi revealed that data he uses are only technical, those relate to establishing new roads, extension or expansion to existing ones or maintenance of old roads. All his work is expressed in technical terms, his "plan" on work is measured in kilometers of roads to be built or maintained in the year and resources needed are also measured in technical resources. This engineer highlights that even translating his budget to financial terms is done not by himself but by an accountant with old experience in the construction field. The result that can be reached here is that for decisions in different industries, there are different types of information needed and that information constitute the body of knowledge that should be captured, organized, disseminated in each sector in both financial and non financial terms. This result agrees with Sabina-Cristina (2007) on knowledge management of some accounting decisions, study applied in Romania. Sabina-Cristina (2007) found that knowledge of making accounting decisions is strictly specific to those decisions, with an emphasize on qualitative knowledge and participants acted based on knowledge in broad context of making decisions not looking at optimization.

In all cases, participants confirm that their organizations have organized databases and well structured files with data about customers, employees, suppliers, costs and prices, payments and revenues collected or charged to customers. Some businesses stress the importance of specific type over others. For instance, in insurance field, a respondent emphasized the importance of regulations and policies saying: *We are tightly regulated field. Prices, premiums, conditions of each policy are by large regulated from the supervising authority. We have a limited margin to "look different" from other companies. Therefore, regulations and guidelines are major part of information we refer to when working.* Unlike this sector, the education representative record his delightfulness that there is lots of demand for academic qualifications. In UAE providers of such qualifications are interested in providing different programs and promote what their institutions have as competitive advantages. The information about the courses and facilities provided, are crucially important. Hotel and leisure is a flexible industry looks for creative ideas and thus has different concerns for knowledge. Respondents from this field considered information about promotion package, prices, quality of services are in the main focus when taking decisions. Participants highlight that "*Dubai particularly became an iconic destination for holidays and many*

hotels are there, so competition is intense. Every place wishes to attract guests by promoting an advantage; it could be price, attraction of destination, variety of services provided and so on. Information about innovative ideas create money, new *ideas are well appreciated in our field"*.

It depends on the type of decision how much financial or non-financial data are used in a decision. It also depends on the nature of the decision whether it is operational or strategic, that detail-related information are used or "overall" information, technical or managerial, are used as with strategic decision like opening new branch or purchase another business. However, users of information highlight a fact that, information, or knowledge they use do not depend only on "raw data" as those stored in databases, it also needs regulations, management decrees and policies, discussion with co-workers in other departments. Very often users face complicated situation, where to resolve, they use: stored data, related regulations or policies and use their own judgment as well. Therefore, there is no inclusive list of information items that relates to a decision.

The author find the participants' feedback generally in line with the literature of accounting information systems (e.g. Romney and Steinbart 2009) where it is decided that there are different information needed for each decision and it depends on the nature and scope of the decision whether the information will be detailed or general, financial only or both financial and non-financial, related to a specific unit or to the overall organization. It is not only the nature and scope of decision but also the type of industry that clearly affect choices and preferences of knowledge to be gathered, maintained and utilized.

Format of Information

When participants were asked about the format of knowledge they use in their decisions, they all respond that the formal body of knowledge is the main one used, comes from databases, policies and regulations, previous business cases and others. Contributors in the experiment reveal that they do not exclude informal information passed by colleagues chat or from the media or any speculations here or there. Hearing a rumor if not taking into account when considering a decision, could rise the level of risk relates to a decision. Nonetheless, evaluating an informal information as strong or weak and influential or not, depends of the recipient's view and his/her attitude towards risk. This result is rather different from the one in the original study by Edward et.al. (2006). In that study, participants placed greater reliance on individual and informal systems although they wished to move more to the formal system. The reason they presented was that the informal knowledge system is rich but the formal is more robust and consistent which enhances shareability.

The author argues here that because the majority of work force (and hence the participants) in UAE are non-locals could be the reason for the higher tendency to rely on formal versus informal and written versus oral information. In this study, it is observed that the vast majority of applicants prefer the written format. One evidence is provided by a participant in food and beverage producing comments: *I make sure that I keep a copy of any email to customer, supplier,...etc or copy of an invoice or purchase order to defend my decision in any case.* Another respondent in municipality added: *If I am in a situation which is new or complicated to me, I use my judgment, I prefer to communicate with my manager about it, however, I take the decision on my own, if I am sure of it, ultimately I do not want to cause any harm to my organization.* The writer considers these views actually reflect sharing of responsibility and different risk-taking more than an organizational culture. Informal information no matter how important, are not evident to be used on a wide scale. This result agrees with the argument of DeLong and Fahy (2000) that culture influences the

behaviors central to knowledge creation, sharing and use. Culture shapes assumptions about what knowledge is, which knowledge to be managed and defines the relationship between individual and organizational knowledge. Moreover, it determines who is expected to control specific knowledge. Contributors in this experiment who are mainly non-UAE citizens, seem to prefer to avoid risk and to justify their decisions based on solid evidence. In their view, that evidence exists mainly in "written" and "formal" information. This relationship maybe mandated by managerial level. Most of the participants belong to middle management from whom the result appears. It may be possible to have a different result if individuals were from senior management with more risk taking.

Source of Information

Respondents were asked if they get their knowledge from internal or/and external sources. All sources are in action in UAE business environment. All respondents report that they have organized internal sources for knowledge including, formal IT systems with databases, documents, emails, regulations and policies, discussions with managers and colleagues. Outside the organization, there are conferences, workshops, exhibitions and other events in addition to magazines, customer surveys, news, words from friends ...etc.

In fact the internal sources is the first and the main source for information for the different knowledge for different situations. However, work environment is known to be a very open environment in terms of organizational learning. Almost all respondents report that generally they "learn" from many sources; formal training courses, workshops and conferences which companies send employees to. Almost all organizations are actively involved in a wide range of events that take place in the country all year through, even if not related directly to their own operations. For example, any small or medium size organization would receive invitations to conferences that may relate directly or indirectly to their work for free. The availability of suck opportunities largely enhance the knowledge of employees and provide big chances to outreach different audience and customer or business partners. Participants from travel agency, hotel and food producers for example mentioned that they attend –at least- the annual international conference about travel and leisure management that is held in Abu Dhabi National Exhibition Center in addition to many others in Dubai. In this field, participants recorded that leisure and hospitality industry is every changeable and developed very fast and it is import to follow up what's new there. Those opportunities allow agents to introduce their companies and their services to customers from all over the world, it is a handy. Emirate of Dubai is particularly very active in that domain.

Regarding technical knowledge, engineer in municipality mentions that his work place invites experts in the field to enrich and refresh their knowledge in the modern methods used in his field. The municipality as a governmental unit believes that investing in people and developing their knowledge will certainly be reflected in the way of doing the work, it may improve quality, reduce time, help attendees to provide new and innovative ideas. Costs for workshops, conferences, specialized events all will pay at the end.

II: How Knowledge Is Managed: How Knowledge Is Acquired, Processed, Shared and Used

All applicants report that their knowledge is gained over time and from experiences they pass in various situations. As mentioned earlier the definition adopted in this research for knowledge is that, the body of items maintained collectively in our minds. It includes data, information, facts, speculations and the new information developed from connected and analyzed information. Some participants referred to this meaning of knowledge in participants' minds. In

addition, respondents recorded that there are formal methods for gathering information in their organizations. Participants mention that "*no organization could live without a memory*". Information and knowledge are stored in databases and organized files. Those form the memory of the organization. It contains information about customers, suppliers, employees, costs, revenues, assets and liabilities and resources owned. Both financial and non-financial information are collected over time and stored. Previous business operations (sales, purchase, investment,....etc) all are part of the body of knowledge of the organization. It includes different types of knowledge, raw data (e.g. contact information of customers, staff,..etc), information (i.e. processed data such as: purchase order executed, computation of net income of a branch or a type of product done before), regulations and policies (those could be in email or statement,...etc). business plan, vision, mission,...etc. For organizations, knowledge is collected in files, documents, databases, emails and informally colleagues discussions or media and publications.

Regarding knowledge processing, contributors in the study found it very hard to explain how they "process" or "use" knowledge in their minds. It is a mental and psychological individual process that differs from one person to another. People could not explain, only psychology and brain-work discipline could tell how this process and the process of utilizing knowledge occur. But participants report that organizations process only formal information and knowledge in specific ways. Regarding accounting decisions, finance officer in retail company mentioned that they use an Oracle package of accounting, where they can not only build up budget but other accounting models such as recording transaction, preparing financial statements and keeping information about customers and purchases and payment and suppliers, sales and collection of payment when inputting data. Similar remark was made by other participants in municipality and other organizations. A participant in insurance with a former experience in police department mentioned that in his previous workplace they were using Peach Tree accounting software. Other programs like the Indian-developed programs were there in practice. The results are consistent with those of Edward et.al. (2006) where found all examined organizations have effective methods of acquiring information but not in sharing and utilizing them and that was everyone's problem not only top management. The latter study also record a difference in view between the organization and employees. Organizations focus on technical solutions concerned largely with making better use of databases and intranet while people are concerned with staff retention, motivation training and networking and hence they wish to have KM systems to do so. They conclude that most (if not all) KM initiatives in organizations will not succeed unless it is backed by people who have access to sufficient resources to make it work. Edward et.al. (2006) make an important conclusion here, if these changes are made, organization would move from being information-based, to intelligence-led and this needs a supportive culture including a strategy/framework for KM.

One note the author makes here is that, all organizations have computerized systems for gathering and storing formal knowledge and for processing routine jobs only. The contrary is also true and by participants' support. Respondents mentioned that strategic decisions are more complicated and while based on both overall financial and non-financial data being analyzed in some ways, they are also extensively discussed between members of higher management and maybe with the help of external expertise. Those decisions include much of management's evaluation and assessment of many factors and maybe guided by similar and other cases in other businesses.

Regarding sharing of knowledge, participants confirm that a healthy organization would promote sharing of knowledge. The way this happens differ from one place to another. In hotels, participants mention that when they face an unusual situation, they enquire from more experienced member who should share their knowledge for the sake of the benefits of the company. In municipality, individuals feel they are more responsible for decisions that affect businesses and individuals in Abu Dhabi (involving in giving a business license, inspections on businesses,...etc). This appreciation of responsibility makes the individuals seek colleagues and senior members advice or guidance when facing uncertainty. Participants from municipality report: "*Individuals involved in a decision can call for a meeting by an email to exchange views with related co-workers and straight away a meeting could be held to discuss the issue of concern*". However, companies avail databases, files, models and format, patterns and previous operations' files and maintaining copies of documents are formal means for sharing knowledge. Those methods represent methods of sharing knowledge. The researcher would regard the result here to be generally in line with O'Leary (2002) who considers that ultimately the value-added from knowledge management can only be known if effective processes are in place to facilitate control, value assessment, and being cost-effective and businesses need some evidence or measurements on a financial return for the use of KM. Measurements may include issues like time spent in collaborating, percentage of group work, extent and complexity of network connecting knowledge.

UAE Environment's Implications on KM

There are some issues from previous findings that highlight business environment in the UAE and its implications on KM. First, regarding knowledge to be captured and managed, participants report that the emphasize which the government places on tourism, raises the importance of collecting and processing information about quality and innovation in that field whereas, the financial services sector generally and insurance specifically is a tightly regulated sector, so information about rules and regulations are more important. Another feature in this country is about the huge inspiration toward getting a higher education degree. This feature motivates education institutions to collect, maintain and distribution more information about degrees and their requirements. Non-local employees in the UAE tend to prefer to depend on written information more than oral which locals feel more confident to use. However, it is evident that the UAE is an ambitious country and achieved a great deal of advances in services to people. This feature is reflected in the existence of computerized and well organized methods of collecting, processing, sharing of information and using by employees from different backgrounds added to the value created via KM systems. Another unique feature of the UAE relates to being a rich oil producer country with small population and a willingness to enhance knowledge and skills of people, it was vast reported that many conferences, workshops and seminars are available which helps in creating and sharing knowledge in different functional areas. These huge resources of companies and sometimes the international requirements which some supervising bodies urge organizations to follow, lead to develop staff standards. This is a feature to be counted for KM practice in the UAE that may not be observed in a country that experience limited resources or struggle under financial crisis. The last feature in business environment in all GCC and the UAE is being multi-culture, it was obvious in this study that having different minds enrich knowledge being created and managed in an organization. It is well known that the UAE is a demanding place in terms of qualifications and experience which work places require to hire staff. This result in enriching amount and quality of the body of knowledge which an organization works with. Such a feature may not be present in a closed society where individuals rarely work with colleagues and managers with different experiences.

Conclusions

This research targets to examine how participants from organizations in UAE acquire, share and utilize knowledge. The findings show that participants do not differentiate between information and knowledge. Results show that all organizations have effective methods of acquiring knowledge in databases, files and other documents and different information are kept and used for the different decisions. Knowledge are processed in computerized programs but how participants could not explain how they process knowledge in their minds and how they utilize in the different situations. The study revealed that there are differences in knowledge and its management according to industry, managerial level and type of decision. there were reflections of the UAE's business environment with its mutli-culture and having specific strategies that encourage tourism industry and tightly regulate insurance. The results of the study are limited by the choice of experiment primarily as a method of investigation and by the number of participants, their experience and the fields they work in. The study was concerned to explore specific points, other elements are missing such as how the knowledge is generated and how used by participants.

References

Benbya, H. (2008). Knowledge Management Systems Implementation: Lessons from the Silicon Valley, *Oxford, Chandos Publishing*.

Bhatt, G. (2001). "Knowledge Management in Organizations: Examining the Interaction between Technologies, Techniques and People," *Journal of Knowledge Management*, Vol. 5, No.1, PP:68-75.

Brikett, W. P. (1995). 'Management Accounting and Knowledge Management,' *Management Accounting*, Nov. PP: 44-48.

Cheng, M. M., Schulz, A. K.- D. & Booth, P. (2009). "Knowledge Transfer in Project Reviews: The Effect of Self-Justification Bias and Moral Hazard," *Accounting and Finance.* Vol. 49 No. 1, P:75.

Chou, T. H., Vassar. J. A. & Lin, B. (2008). "Knowledge Management via Ontology Development in Accounting," *Kybernetes, The international journal of cybernetics, systems and management sciences,* Vol. 37 Iss: 1, pp.36 – 48.

Christensen, L. B., Johnson, R. B. & Turner, L. (2011). Research Methods, Design and Analysis, 11 ed., *Pearson.*

Courtney, J. F. (2001). "Decision Making ad Knowledge Management in Inquiring Organizations: Toward a New Decision-Making Paradign for DSS," *Decisions Support Systems,* Vol. 31, PP: 17-38.

De Long, D. W. & Fahy, L. (2000). "Diagnosing Cultural Barriers to Knowledge Management," *Academy of Management Executive,* Vol. 14, No. 4, PP: 113-127.

Edwards, S. J., Collier, P. M. & Shaw, D. (2006). 'Knowledge Management and its impact on the Management Accountant,' *Research Report for CIMA, UK.*

Firestone, J. (2001). 'Key Issues in Management Knowledge,' *Knowledge and Innovation: Journal of the KMCI,* Vol.1 No.3.

Kaplan, R. S. & Johnson, H. T. (1987). 'Relevance Lost: The Rise and Fall of Management Accounting,' *Harvard Business School Press.*

Maier, R. (2007). "Knowledge Management Systems: Information And Communication Technologies for Knowledge Management," 3rd edition, Berlin, *Springer.*
National Bureau of Statistics in UAE (2010).   ,
http://www.uaestatistics.gov.ae/ReportDeta ilsEnglish/tabid/121/Default.aspx?ItemId=1 923&PTID=104&MenuId=1, as checked in 20/2/2012.

O'Leary, D. E. (2002). "Knowledge Management in Accounting and Professional Services," in Researching Accounting As an Information System Discipline by Vicky Arnold and Steve G. Sutton, Published by *American Accounting Association.*

Roberts, H. & Bhimani, A. (2004). "Management Accounting and Knowledge Management: In Search of Intelligibility," *Accounting research,* Vol. 15, No. 1. PP: 1-4.

Romney, M. & Steinbart, P. (2009). 'Accounting Information Systems,' 11th Edition, *Prentice Hall.*

Sabina-Cristina, M. (2007). "The Accounting Decisions and their Modelling by Using Specialized Computer-Based Tools," *The International Journal of Digital Accounting Research,* Vol.7, No.12, PP: 27-51.

Sori, Z. M. (2009). "Accounting Information Systems (AIS) and Knowledge Management: A Case Study", *American Journal of Scientific Research,* Vol. 4, pp36-44.

Taylor, E. Z. (2006). "The Effect of Incentives on Knowledge Sharing in Computer-Mediated Communication: An Experimental Investigation," *Journal of Information Systems,* Vo. 20, No.1, PP: 103-120.

Thomas, R. M. (2003). "Blending Qualitative & Quantitative Research Methods in Theses and Dissertations," *Sage Publication, New Orleans.*

Wilkinson, J. (2000). 'Accounting Information System,' 7th Ed. *John Wiley & Son.*

13

A Framework of Critical Factors to Knowledge Workers' Adoption of Inter-organizational Knowledge Sharing Systems

Kamla Ali Al-Busaidi

Sultan Qaboos University, Alkhod, Oman

Academic Editor: Haslindar Ibrahim

Abstract

The objective of this research is to outline the critical factors to knowledge workers' adoption of "public-good" Inter-organizational knowledge sharing systems (IOKSS). Public-good IOKSS is one that is open to all firms in a specific sector. Public-good IOKSS is valuable and critical for effective social and economic development of any sector, especially knowledge-based sectors. However, the deployment of inter-organizational knowledge networks and collaborations may incur costs, challenges and risks to organizations and their individuals. To deploy a sustained inter-organizational electronic knowledge sharing, these obstacles must be overcome. Prior knowledge sharing research has mostly focused on the within-firm context. Prior research on inter-organizational systems (IOS) has focused on organizational adoption, and most empirical studies have mainly focused on supply-chain organizations, those that are vertically-linked. Based on KM and IOS literature, the critical factors to knowledge workers' adoption of public good IOKSS can be related to individual factors, their relationships with peers, the organization, the proposed IOKSS system, and the sector.

Keywords: Knowledge Workers, Inter-organizational System, Inter-organizational Knowledge Sharing System, Theoretical model.

Introduction

Knowledge is a powerful resource that enables nations, organizations and individuals to achieve several benefits such as improved learning, innovation and decision-making. Any organization, public or private, requires knowledge management to achieve its best performance. A Knowledge Management System (KMS) "is an integration of technologies and mechanisms that are developed to support knowledge management processes" (Becerra-Fernanadez et al., 2004, p. 31). An Inter-organizational KSS (IOKSS) is a type of KMS and is defined here as a system that enables seamless dissemination of individual and organizational knowledge (through repositories or networking) between two or more organizations. This study aims to investigate the key success factors for the development of "public good" IOKSS among organizations at the same business level (horizontally-linked) in a specific nation. A "public good" IOKSS is one that is open to all firms even if they did not contribute to the development of the system (Choudhury, 2007).

An IOKSS is one type of inter-organizational system (IOS). The concept of an IOS was coined originally by Cash and Konsynski (1985) and defined as an automated information system shared by two or more organizations, and designed to link business processes (Cash and Konsynski, 1985; Robey, et al., 2008). An IOS can result in several operational, strategic and social benefits for the participating organizations (Barrett and Konsynski, 1982; Robey et al., 2008), the government and society.

There is a persistent need for developing nations to take advantage of new technologies for acquiring and disseminating knowledge. Inter-organizational information integration is crucial for digital government. Partnerships between public and private organizations in specific sectors, especially service or knowledge-based sectors, are vital for the social and economic development of any country and the welfare of society such as in health and education sectors. For instance, an IOKSS can be developed for the health sector to enable physicians in the same or different organizations in the health sector (hospitals, medical centers, etc.) to share knowledge (medical cases, treatments, medical reports). Physicians can share knowledge through health IOKSS by either codifying it in the system, or communicating the knowledge with colleagues in other hospitals through the system (e.g., using video-conferencing). It fosters training and learning among knowledge workers, and lessens the knowledge gap among professionals. Moreover, such initiatives

provide support for the government's developmental decision making and planning.

However, there are other costs and risks to establishing inter-organizational networks and collaborations (Williams, 2005). These risks and barriers are linked to individual, organizational, technological, social and political factors related to different stakeholders including for those organizations in horizontal linkages. Knowledge sharing is a challenging process even within a specific organization, as not many people are willing to share their best practices. There are a number of studies that have investigated the enablers (or motivators) of general knowledge sharing behaviour in an organizational context such as Bock et al. (2005), Kankanhalli et al. (2005), Wasko and Faraj (2005), Al-Alawi et al. (2007), Al-Busaidi et al. (2010) and Chen et al. (2012). However, crossing the boundary of an organization through IOKSS will additionally complicate the knowledge sharing process. These obstacles must be overcome, to develop smooth and sustained inter-organizational networks, including IOKSS.

There are several theoretical non-empirical papers on the development of inter-organizational systems (IOS) such as those of Barrett and Konsynski (1982), Cash and Konsynski (1985), Boonstra and De Vries (2005), and Robey et al. (2008). However, most of the prior papers reported empirical studies mainly focused on organizational adoption (Bala and Venkatesh, 2007; Robey et al., 2008), but gave inadequate attention to the context of these IOS (Makipaa, 2006). Very limited studies have assessed IOKSS adoption by knowledge workers, who are the end users, and therefore key stakeholders for achieving the expected benefits in a knowledge management initiative. In a knowledge-based system, knowledge workers, the end users, are the driving force of the system; such a system can only survive through their commitment and use. Thus, recognizing knowledge workers' attitude at a very early stage will enable the organization to make a better decision and ensure end users' commitment throughout the whole development process.

IOKSS can be deployed to connect organizations in vertical linkage such as those in the supply chain (suppliers, organizations, customers, etc.), or organizations in horizontal linkage (those that operate in the same business level). Most prior empirical studies have investigated IOS deployment in vertical linkage of organizations (Reich and Benbasat, 1990; Grover, 1993; Rai et al., 2006; Ranganathan et al., 2011). However, a few (e.g., Pardo et al. (2006); Dawes et al. (2009); Yang and Maxwell (2011)) address IOS adoption for organizations that participate in horizontal linkages. Linking rivals through IOKSS can be more challenging than linking organizations in vertical linkage because of competition and rivalry, especially if it involves private organizations. With respect to horizontal linkage, some researchers who address the public sector have developed theoretical frameworks for IOS deployment such as Pardo et al. (2006), Dawes et al. (2009) and Yang and Maxwell (2011).

Organizations that are horizontally-linked can develop IOS cooperatively for strategic alliance or/and public good (Choudhury, 1997). Little is empirically known about the enablers of sharing knowledge in systems that connect organizations (public or private) in horizontal linkage in a specific sector or industry. Knowledge sharing processes and systems will not only be challenged by individual and organizational factors but also can be challenged by social, technical and political inter-organizational factors.

Determining the critical factors that impact knowledge workers' adoption (attitudes and intentions) of IOKSS in horizontally-linked organizations will help to improve organizational adoption of such systems. The goal of this research is to outline the critical factors to knowledge workers' adoption of IOKSS in horizontally-linked organizations.

Based on KM and IOS literature, these factors include the characteristics of the knowledge workers and managers, their relationships with peers, the organizations, the proposed system (IOKSS), and the sector/industry.

IOKSS Success Factors

Knowledge Workers Adoption

As previously mentioned, this study aims to develop an explanatory model of the antecedents of knowledge workers adopting of inter-organizational knowledge sharing systems (IOKSS) for horizontally-linked organizations in service-based sectors that include public and private organizations. The antecedents of IOKSS are related to the factors that produce effective knowledge sharing behaviour and boundary spanning. Thus, the antecedents of IOKSS may be related to individual (computer self-efficacy, personal innovativeness, knowledge self-efficacy, image, knowledge ownership perception, perceived benefits/costs), peers (attitude, interactivity level, trustworthiness), system (perceived ease of use, perceived usefulness, perceived security, perceived compatibility), organization (management support, rewards, technological competence, organization structure), and sector (support, regulations/policies, standardization level, competitive pressure, information systems homogeneity). These factors are shown in Figure 1.

Previous studies have investigated knowledge sharing based on: (1) attitude as in Chen et al. (2012) and Jeon et al. (2011); (2) intention as in Chen et al. (2012), Jeon et al. (2011) and Bock et al. (2005); or actual knowledge sharing behaviour as in Kankanhalli et al. (2005), Wasko and Faraj (2005), Al-Busaidi et al. (2010) and Alawi et al. (2007). Since this model assesses knowledge workers' perception of IOKSS at the pre-implementation stage, knowledge sharing will be assessed by knowledge workers' attitudes toward the systems and intention to adopt IOKSS in general and to share knowledge.

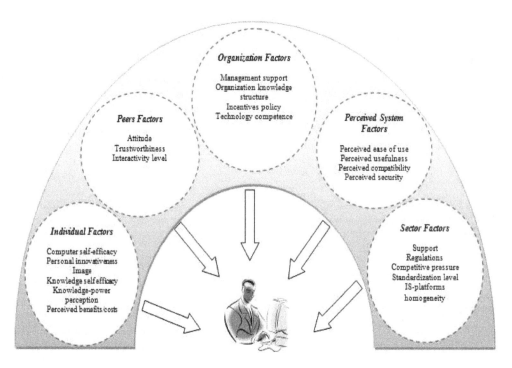

Fig 1. Critical Factors to Knowledge Workers' Adoption of "Public Good" IOKSS

Individual Factors

There are several individual factors that might affect knowledge workers adoption of IOKSS; these factors can be related to their self-efficacy, personal innovativeness, knowledge self-efficacy, image, knowledge ownership perception and perceived benefits/costs.

Knowledge workers' computer self-efficacy may impact their perception of new technological innovations such as IOKSS. "People's judgments of their capabilities to organize and execute courses of action required to attain designated types of performances" (Bandura, 1986, p.391). In the knowledge-based systems context, self-efficacy significantly impacts knowledge contributors' usage of the system (Kankanhalli et al., 2005). Not many IOS studies have investigated the impact of end users' characteristics. However, having sufficient technical skills to use IOKSS, will

reduce the resistance to change factor that is often highlighted as a barrier in IS research including KMS.

Knowledge workers' personal innovativeness may also impact their attitude toward IOKSS. "Being used to adapting to new systems and processes might reveal the usefulness and ease of use more quickly to an innovative person than to a non-innovative person" (Schillewaert et al., 2005, p.843). Personal innovativeness influences end users' adoption of a knowledge management system (Xu and Quaddus, 2007). Also, willingness to experiment may impact adoption of inter-organizational knowledge networks (Dawes et al., 2009).

Knowledge self-efficacy is critical to individuals' knowledge sharing behavior especially in electronic format including IOKSS. Professionals with high expertise may feel confident to advice and consequently share their knowledge;

whereas those who have insufficient knowledge may feel incompetent and are reluctant to share their knowledge. Knowledge-self efficacy is significant knowledge contributors' usage KMS (Kankanhalli et al., 2005).

Individuals' perception of knowledge as power may impact their adoption of IOKSS. Knowledge is perceived as power by many individuals especially in knowledge-based services, and this perception typically affects knowledge sharing behavior (Wang and Noe, 2010). Thus, individuals may be reluctant to share their knowledge as they may feel it would reduce their value (power). Yang and Maxwell (2011) indicated that information/knowledge ownership can be a critical factor of information sharing through IOS in public organizations, especially if it is not part of the organizational culture.

Image can be an important social factor that affects knowledge workers' attitude and intention toward IOKSS in their domain. Image is the degree to which an individual believes the use of an innovation will improve one's position in one's social system (Moore and Benbasat, 1991). Bock et al. (2005) found that subjective norm, which is positively associated with image, impacts individuals' intention to share knowledge. In the IOS context, Yang and Maxwell (2011) indicated that people are motivated to contribute to the collective good in an organization as long as they maintain their social identity.

Perceived benefit/cost can be a major issue in individuals' attitude and decision to use IOKSS. Perceived benefits/costs impacts individuals' knowledge sharing behavior (Wang and Noe, 2010). Knowledge sharing can results in benefits such as enhancing professional reputation Wasko and Faraj, 2005), and enjoyment in helping others(Kankanhalli et al., 2005). There are some perceived costs about knowledge sharing (such as time, efforts and loss of power) that might negatively impacts individuals' attitude toward it.   In the IOS

context, direct and indirect perceived benefits and costs can be major factors in IOS adoption (Robey et al., 2008; Yang and Maxwell, 2011).

Peers Factors

There are several peers' factors that might affect knowledge workers adoption of IOKSS; these factors can be related to peers' attitude, trustworthiness and interactivity. Based on social influence theory (i.e., subjective norms), peers' attitudes toward the technology may impact individuals' attitudes toward the technology (Ajzen, 1991) and technology acceptance (Venkatesh and Davis, 2000). In the knowledge-based system context, Bock et al. (2005) found that subjective norm affect individuals' intentions to share knowledge. In public good IOKSS, individuals may have positive attitudes toward IOKSS and knowledge sharing if they think that their peers in the domain have knowledge/information needs. Partner's commitment affects the extent to which companies undertake IOS integration and increase the percentage of IOS exchange and performance (Lee and Lim, 2005). This principle can be also applied at the knowledge workers level.

Knowledge sharing or "selling" in an organization depends on the trustworthiness of the knowledge utilizers (or buyers) (Davenport and Prusak, 1998). The significance of trust in several knowledge-based activities including knowledge externalization was found to be statistically significant (Lee and Choi, 2003). Al-Alawi et al. (2007) found that trust is positively related to knowledge sharing in organizations. Lee and Lim (2005) found that a partner's trust affects organization adoption of IOS and the effectiveness of the IOS' performance. This principle can be also applied at the knowledge workers level because if individuals trust their peers within their organization and across organizations in this integration of knowledge sharing, it can positively impact their performance.

Peers' existing inter-organizational communication and social networking level in a specific domain may impact the individuals' attitudes toward inter-organizational knowledge sharing behaviour. The level and quality of interaction among peers in specific domains illustrates the need for IOKSS and motivates its implementation. Team cohesiveness and communication style were found to be positively related to individuals' knowledge sharing behaviour (Wang and Noe, 2010). Yang and Maxwell (2011) indicated that the existing social network impacts intra-organizational information sharing, including inter-organizational information sharing in the public sector.

Perceived System Factors

Several IOKSS factors may contribute to knowledge workers' adoption of IOKSS. These factors can be related to IOKSS' perceived ease of use, perceived usefulness, perceived compatibility and perceived security. Perceived ease of use can be a factor in knowledge workers' perception of IOKSS. Researchers indicate that perceived ease of use affects users' intention to use a technology (Bailey and Pearson, 1983; Venkatesh and Davis, 2000). Having a user-friendly, easy to learn and use knowledge management system influences end users' adoption (Xu and Quaddus, 2007). Likewise, perceived ease of use of IOS can be a critical factor to its implementation (Yang and Maxwell, 2011).

Perceived usefulness and capabilities can impact knowledge workers' perception of any information system including IOKSS. Perceived usefulness is one of the main significant factors on individuals' acceptance of a technology (Venkatesh and Davis, 2000). Perceived usefulness was found a significant factor on professionals' attitude toward knowledge sharing (Hung et al., 2010). Perceived usefulness of IOS can be a critical factor to its implementation (Yang and Maxwell, 2011).

Perceived compatibility of IOKSS with work practices is critical to knowledge workers' attitude toward the system. Perceived compatibility (how consistent is the innovation with an individual's values and experience) is an important characteristic of KMS to end users' adoption (Xu and Quaddus, 2007). Based on Venkatesh and Davis's (2000) technology acceptance model (TAM2), people perceive the system's usefulness by cognitively comparing its capabilities with what they need to get done in their job (job relevance). Based on task-technology fit research, job relevance is an important influence on the acceptance of a technology (Goodhue, 1995). The compatibility of technology is also a factor in IOS adoption and diffusion (Robey et al., 2008).

Perceived security is an important technical issue for the adoption of IOKSS especially when confidential information/knowledge is shared among several organizations. The importance of security on the use of information systems is confirmed by several researchers such as Chang and Wang (2011), and KM researchers such as Gold et al. (2001) and Jennex and Zyngier (2007). Similarly, the importance of security on IOS is highlighted by Boonstra and De Vries (2005), Suomi (1993), and Yang and Maxwell (2011). Designing IOS with access authorization and authentication is critical for sharing information (Suomi, 1993; Yang and Maxwell, 2011).

Organization Factors

Organization factors such as management support, organization structure, incentives policy and technology competence might impact knowledge workers adoption of IOKSS.

Management support for knowledge exchange reduces individual experts' fear of losing their values. Management support is critical for KMS (Davenport and Prusak, 1998; Gold et al., 2001), and extremely critical to endorse KMS including the IOKSS, and consequently change employees' attitudes. Management support was found to be significantly correlated with knowledge sharing behaviour (Al-Busaidi et al., 2010). IOS literature has also emphasized the importance of top management support for IOS adoption (Grover, 1993; Robey et al. 2008).

Organizational structure is essential in leveraging technological architecture and knowledge management effectiveness (Gold et al, 2001; O'Dell and Grayson, 1998). Creating a flexible organizational structure that endorses knowledge sharing within an organization optimizes knowledge sharing not only within the organization but also across the organization's boundaries (Gold et al., 2001). The importance of the organization's structure s is also highlighted by Robey et al. (2008) and Yang and Maxwell (2011).

Organizations that want to encourage and promote employee interaction and knowledge sharing activities should adopt an incentives/rewards policy (Davenport and Prusak, 1998). Rewards policy is positively related to knowledge sharing in organizations (Al-Alawi et al., 2007; Al-Busaidi et al., 2010). Rewards/Incentives for the implementation of IOS are highlighted by Robey et al. (2008) and Yang and Maxwell (2011). Without good incentives employees will be reluctant to exchange and contribute their own knowledge to the KMS (O'Dell and Grayson, 1998).

The technical systems and infrastructure within an organization can contribute to knowledge workers' attitude toward IOKSS. A technological infrastructure that supports the communication of various types of knowledge is critical for building a firms' knowledge infrastructure capability (Gold et al., 2001) and the development of IOS (Lin, 2006; Robey et al., 2008; Yang and Maxwell, 2011). Having a compatible IT infrastructure (good technology and competent IT staff competency) is a major enabler of IOS, and

improves knowledge workers' attitude and adoption of IOKSS.

Sector Factors

Several factors related to the sector where IOKSS is deployed might impact knowledge workers' adoption of IOKSS. These factors include sector support, regulations, competitive pressure, standardization level and homogeneity of organizational IS-platforms in different organizations in the sector.

Sector/government support for a "public good" IOKSS can be positively associated with knowledge workers' attitude toward its implementation and adoption. If a sector or related government agency fully supports such an initiative, knowledge workers should value it and consider it useful. This positive impact can be also explained by Ajzen's (1991) social influence theory which suggests that if one's superior thinks using a system is useful, then that person may also believe it. Sector or government pressure impacts IOS adoption and diffusion (Robey et al., 2008). For instance, Dawes et al. (2009) indicated that acquiring legal authority through an existing statute for a knowledge network in the public sector is a necessity.

Regulations and policies can be a major factor impacting knowledge workers' adoption of IOKSS. Regulations may hinder the adoption of IOKSS because government may prohibit sharing sensitive and regulated information in domains and sectors related to public safety and national security (Pardo et al., 2006; Yang and Maxwell, 2011). Legal barriers related to cross-organizational information transfer inhibit IOS adoption (Boonstra and De Vries, 2005; Robey et al., 2008). Having government/industry regulations that support IOS deployment and knowledge sharing across organizations in a specific domain encourages individuals to contribute to the systems and have positive attitudes toward its deployment.

Competitive pressure in a specific sector may also impact knowledge workers' attitude toward IOKSS implementation. Competitive pressure results from a threat of losing competitive advantage and therefore positively affects IOS planning effectiveness (Lin, 2006). Likewise, IOS research has highlighted the impact of competitive pressure on organizational adoption of the system. Lin (2006) found it significantly correlated with internet-based IOS adoption.

Having standardized business processes and shared understanding among firms in a specific sector can promote knowledge workers' attitude toward IOKSS in that domain. Having standardized business processes and practices can reflect a sense of integration and shared practices in the domain and improve the acceptance of IOKSS. Effective knowledge sharing depends on shared understandings, professional norms and standardized business processes and practices (Dawes et al., 2009). Different operational procedures, workflows and control mechanisms can impact inter-organizational information sharing (Yang and Maxwell, 2011).

Having existing homogeneous IS platforms can enhance knowledge workers' attitude toward IOKSS. Deploying IOKSS in a sector with organizations that use different information systems can negatively impact knowledge workers' attitude toward IOKSS deployment. That is, heterogeneous information systems with different platforms (hardware and software) and data standards can challenge IOS adoption (Boonstra and De Vries, 2005; Mäkipää, 2006; Pardo et al., 2006; Yang and Maxwell, 2011).

Conclusion

Inter-organizational knowledge sharing systems (IOKSS) is very valuable especially for service-based and knowledge-based sectors in any nation. IOKSS can be an integral part of e-government. However, costs, challenges and risks to organizations and their individuals may result from inter-

organizational knowledge networks and collaborations. These obstacles must be overcome to develop stable and sustained inter-organizational networks and inter-organizational electronic knowledge sharing. The majority of prior theoretical and empirical research on knowledge sharing has focused on the within-firm context. In addition, studies on inter-organizational systems (IOS) lacks investigations on knowledge workers' adoption of such systems. Most empirical IOS studies have mainly focused on organizational adoption of IOS in vertically-linked supply-chain organizations. The objective of this research is to develop a model of the key antecedents of knowledge workers' adoption of IOKSS in organizations that are horizontally-linked.

Based on KM and IOS, this study proposed that the antecedents of knowledge workers' adoption of "Public good" IOKSS in a specific sector are related to personal factors, peers factors, organization factors, system factors and sector factors.

This study proposed a detailed framework that can be used by researchers and practitioners to examine knowledge workers' attitude and adoption of IOKSS, and ensure successful deployment of IOKSS. This study only proposed a theoretical model, thus empirical investigations are also needed to verify the effects of these factors. Also, future research should develop or adopt reliable and valid measurements for researcher and practitioners to evaluate this proposed model. Future qualitative studies (such as case analysis, interviews etc.) might reveal some further insights on these factors. However, further quantitative rigorous studies are needed to validate the model and generalize it.

Acknowledgment

This manuscript is a part of a granted research project by Sultan Qaboos University (SQU). The manuscript also benefited from the feedback received from Professor Lorne

Olfman at Claremont Graduate University in California.

References

Ajzen, I. (1991). "The Theory of Planned Behavior," *Organizational Behavior and Human, Decision Process,* 50 (2), 179-211.

Al-Alawi, A. I., Al-Marzooqi, N. Y. & Mohammed, Y. F. (2007). "Organizational Culture and Knowledge Sharing: Critical Success Factors," *Journal of Knowledge Management,* 11(2), 22-42.

Al-Busaidi, K. A., Olfman, L., Ryan, T. & Leroy, G. (2010). "Sharing Knowledge to a Knowledge Management System: Examining the Motivators and the Benefits in an Omani Organization," *Journal of Organizational Knowledge Management.* [Online], [Retrieved September 2012], http://www.ibimapublishing.com/journals/J OKM/2010/325835/325835.html

Bailey, J. E. & Pearson, S. W. (1983). "Development of a Tool for Measuring and Analyzing Computer User Satisfaction," *Management Science,* 29(5), 530-545.

Bala, H. & Venkatesh, V. (2007). "Assimilation of Interorganizational Business Process Standards," *Information Systems Research,* 18 (3), 340-362

Barrett, S. & Konsynski, B. (1982). "Inter-Organizational Information Sharing Systems," *MIS Quarterly,* 6(4), 93–105.

Becerra-Fernandez, I., Gonzalez, A. & Sabherwal, R. (2004). 'Knowledge Management,' *Pearson Education Inc, New Jersey, NJ, USA.*

Bock, G. W., Zmud, R. W., Kim, Y. G. & Lee, J. N. (2005). "Behavioral Intention Formation in Knowledge Sharing: Examining the Roles of Extrinsic Motivators, Social-Psychological Forces, and Organizational Climate," *MIS Quarterly,* 29(1), 87–111.

Boonstra, A. & De Vries, J. (2005). "Analyzing Inter-Organizational Systems from a Power and Interest Perspective," *International Journal of Information Management,* 25(6), 485–501

Cash, J. I. & Konsynski, B. R. (1985). 'IS Redraws Competitive Boundaries,' *Harvard Business Review,* March-April, 134-142.

Chang, K.- C., Wang, C.- P. (2011). "Information Systems Resources and Information Security," *Information Systems Frontiers,* 13 (4), 579-593.

Chen, S.- S, Chuang, Y.- W., Chen, P.- Y. (2012). "Behavioral Intention Formation in Knowledge Sharing: Examining the Roles of KMS Quality, KMS Self-Efficacy, and Organizational Climate," *Knowledge-Based Systems,* 31, 106-118.

Choudhury, V. (1997). "Strategic Choices in the Development of Interorganizational Information Systems," *Information Systems Research,* 8(1), 1-24.

Davenport, T. H. & Prusak, L. (1998). Working Knowledge: How Organizations Manage What they Know, *Harvard Business School Press,* Boston, MA, USA.

Dawes, S. S., Cresswell, A. M. & Pardo, T. A. (2009). "From "Need to Know" to "Need to Share": Tangled Problems, Information Boundaries, and the Building of Public Sector Knowledge Networks," *Public Administration Review,* 69(3), 392-402.

Gold, A. H., Malhotra, A. & Segars, A. H. (2001). "Knowledge Management: An Organizational Capabilities Perspective," *Journal of Management Information Systems,* 18(1), 185- 214.

Goodhue, D. L. (1995). "Understanding the Linkage between User Evaluations of Systems and the Underlying Systems," *Management Science,* 41(12), 1827-1844

Grover, V. (1993). "An Empirically Derived Model for Adoption of Customer-Based Inter-Organizational Systems," *Decision Sciences,* 24(3), 603–640.

Jarvenpaa, S. L. & Staples, D. S. (2001). "Exploring Perceptions of Organizational Ownership of Information and Expertise," *Journal of Management Information Systems,* 18(1), 151–183.

Jennex, M. E. & Zyngier, S. (2007). "Security as a Contributor to Knowledge Management Success," *Information Systems Frontiers,* 9 (5), 493-504.

Jeon, S., Kim, Y.- G. & Koh, J. (2011). "An Integrative Model for Knowledge Sharing in Communities-of-Practice," *Journal of Knowledge Management,* 15 (2), 251-269.

Kankanhalli, A., Tan, B. C. Y. & Wei, K. K. (2005). "Contributing Knowledge to Electronic Knowledge Repositories: An Empirical Investigation," *MIS Quarterly,* 29(1), 113–143.

Lee, H. & Choi, B. (2003). "Knowledge Management Enablers, Processes and Organizational Performance: An Integrance View and Empirical Examination," *Journal of Management Information Systems,* 20(1), 179-228.

Lee, S. & Lim, G. G. (2005). "The Impact of Partnership Attributes on EDI Implementation Success," *Information & Management,* 42, 503–516.

Lin, H.- F. (2006). "Interorganizational and Organizational Determinants of Planning Effectiveness for Internet-Based Interorganizational Systems," *Information and Management,* 43 (4), 423-433.

Mäkipää, M. (2006). "Inter-Organizational Information Systems in Cooperative Inter-Organizational Relationships: Study of the Factors Influencing to Success," *IFIP International Federation for Information Processing,* 226, 68-81.

Moore, G. C. & Benbasat, I. (1991). "Development of an Instrument to Measure the Perceptions of Adopting an Information Technology Innovation," *Information Systems Research,* 2(3), 173-191.

O'Dell, C. & Grayson, C. (1998). "If Only We Knew What We Know: Identification and Transfer of Internal Best Practices," *California Management Review,* 40(3), 154-174.

Pardo, T. A., Cresswell, A. M., Thompson, F. & Zhang, J. (2006). "Knowledge Sharing in Cross-Boundary Information System Development in the Public Sector," *Information Technology and Management,* 7(4), 293–313.

Rai, A., Patnayakuni, R. & Patnayakuni, N. (2006). "Firm Performance Impacts of Digitally-Enabled Supply Chain Integration Capabilities," *MIS Quarterly,* 30(2), 225–246.

Rainer, R. K., Turban, E. & Potter, R. E. (2007). Introduction to Information Systems: Supporting and Transforming Business, *John Wiley & Sons,* Inc, NJ, USA.

Ranganathan, C., Teo, T. S. H. & Dhaliwal, J. (2011). "Web-Enabled Supply Chain Management: Key Antecedents and Performance Impacts," *International Journal of Information Management,* 31(6), 533–545.

Reich, B. H. & Benbasat, I. (1990). "An Empirical Investigation of Factors Influencing the Success of Customer Oriented Information Systems," *Information Systems Research,* 1 (3), 325-347.

Robey, D., Im, G. & Wareham, J. D. (2008). "Theoretical Foundations of Empirical Research on Interorganizational Systems: Assessing Past Contributions and Guiding Future Directions," *Journal of the Association for Information Systems,* 9(9), 497-518.

Schillewaert, N., Ahearne, M. J., Frambach, R. T. & Moenaert, R. K. (2005). "The Adoption of Information Technology in the Sales Force,"

Industrial Marketing Management, 34(4), 323– 336.

Suomi, R. (1993). "What to Take into Account When Building an Inter-Organizational Information System," *Information & Management,* 30(1), 151-159.

Venkatesh, V. & Davis, F. D. (2000). "A Theoretical Extension of the Technology Acceptance Model: Four Longitudinal," *Management Science,* 46(2), 186-204.

Wang, S. & Noe, R. A. (2010). "Knowledge Sharing: A Review and Directions for Future Research," *Human Resource Management Review,* 20, 115–131.

Wasko, M. M. & Faraj, S. (2005). "Why Should I Share? Examining Social Capital and Knowledge Contribution in Electronic Networks of Practice," *MIS Quarterly,* 29(1), 35-57.

Williams, T. (2005). "Cooperation by Design: Structure and Cooperation in Inter-Organizational Networks," *Journal of Business Research,* 58(2), 223-31.

Xu, J. & Quaddus, M. (2007). "Exploring the Factors Influencing End Users' Acceptance of Knowledge Management Systems: Development of a Research Model of Adoption and Continued Use," *Journal of Organizational and End User Computing,* 19(4), 57-79.

Yang, T.- M. & Maxwell, T. A. (2011). "Information-Sharing in Public Organizations: A Literature Review of Interpersonal, Intra-Organizational and Inter-Organizational Success Factors," *Government Information Quarterly,* 28(2), 164–175.

14

Syllabus Management System for Academics Practicing Knowledge Management

Anushia Chelvarayan[1], Chandrika Mohd Jayothisa[1], Hazlaili Hashim[1] and Khairol Nizat Lajis[2]

[1]Centre For Diploma Programme, Multimedia Universiti, Jalan Ayer Keroh Lama, Melaka

[2] Foundation Studies and Extension Education, Multimedia Universiti, Jalan Ayer Keroh Lama, Melaka.

Abstract

Knowledge management, a very popular term which describes a range of practices used by organizations to identify, create, represent, and distribute knowledge for reuse, awareness and learning across the organization. This paper will discuss a practice used by academics in Centre for Diploma Programme (CDP) in Multimedia University via syllabus management system. Every two to five years, CDP's academic and management staff will prepare the updated syllabi for MQA accreditation. The system will assist both the administrator and the lecturers in organizing, updating and retrieving their syllabi information for the entire Diploma in Information Technology (DIT) programme. There are basically a few factors that encourage academics to practice knowledge management. First of all, we will identify the contributing factors that encourage the academics in CDP, MMU for practicing knowledge management i.e. MQA's requirement, academics' commitment and the university's requirement and then the practice of knowledge management is shared through the Syllabus Management System. The requirement models have been represented using the Unified Modeling Language (UML) and the development stage uses the ontology development methodology. The ontology methodology is then used as a guideline for creating ontologies based on a declarative knowledge representation system. The system is deployed on the Protégé ontology editor tool.

Keywords: Knowledge Management, Outcome Based Education, Protégé , Ontology

Introduction

Knowledge is neither equal to data nor information. In fact, knowledge can be described as something that makes both data and information manageable. For example, if you want to travel by train from Rawang to Ipoh, you will need some data, some information and above all, knowledge. You have the data through tables with train times at the Rawang station. From this data, you can extract meaningful and useable information from the large amount of data that are relevant for the trip. The thing that makes this all possible is knowledge. This is because, you have knowledge of train tables and you must consult the tables if you need to know what time your train leaves. You can also read and you found the station. All these things have something to do with knowledge.

Knowledge is characterized by information, a capacity and an attitude. Knowledge management needs to take into consideration the system-bound side of knowledge or also known as information and people-bound side of knowledge or also known as capacity and attitude. The system-bound side of knowledge is called explicit knowledge and the people-bound side knowledge is known as implicit or tacit knowledge.

Knowledge Management (KM) has managed to become the main source and continued key factor in developing and implementing competitive and successful systems that represents the organization memory in various fields including education. This competitive advantage is achieved through the process of creating, collecting, organizing, diffusing and implementing of both creative and timely business solutions that are able to pursuit the organizational objectives.

Multimedia University (MMU) is currently having more than ten faculties and departments offering different education programmes which are accredited by Malaysian Qualification Assurance (MQA) previously known as Lembaga Akreditasi Negara (LAN). To maintain the quality of programmes taught, MQA has provides some guidelines and procedures on syllabus format and materials that need to be collected, prepared and updated. These are to ensure that all programmes offered are always met with its quality assurance.

A Document Management System allows users to track and manage documents across work groups which include handling critical information assets such as lecturer's teaching background, subject learning outcomes and programme outcomes. In order to manage the syllabus, a simple prototype is developed to enable authorized user groups to locate, update, store and retrieve data in the most efficient manner. The system is developed using the Protégé 2000, an ontology editor tool that will assist in defining and providing an extensible architecture of creating and

customizing knowledge based application, in this study the knowledge of syllabi.

Research Objective

The overall goal of this study is to identify the contributing factors that encourage knowledge management practice among academics via the syllabus management system in Centre for Diploma Programme, Multimedia University Melaka. Moreover, we need to identify if MQA requirement, academics' commitment and university requirement are the contributing factors that encourage knowledge management practice among academic in CDP, MMU. Based on the above research objectives, a syllabus management system is introduced to encourage knowledge management practice among the academics. A simple prototype has been developed and introduced to the academics by using an ontology tool in identifying the knowledge base and representing the organization memory by emphasizing on the knowledge of programmes. A simple prototype is developed based on the models, structure and the defined knowledge base repository of the Syllabus Management System.

Scope and Limitations of Study

The scope of this research is as follows: (1); The contributing factors for knowledge management practice among academic is only limited to three factors that is MQA's requirement, academics' commitment and university's requirement, (2) The Document Management System only caters for the Diploma in Information Technology syllabi, (3); The system is able to store, update, search and retrieve syllabus. (4); The use of ontology tools to module the Syllabus Management System

Literature Review

Knowledge Management (KM)

KM has managed to become the main success factor for organizations in building and representing their organization memory (OM) in various fields such as education, engineering, management and environment. Information Technology (IT) has proven to continuously support KM throughout the development of OM. In addition, the integration of informal, semiformal, and formal knowledge helps to facilitate its access, sharing and reused by the members of their organization(s) for solving their individual or collective tasks (Thorsten & Sure, 2002). A common approach to tackle the knowledge management problem in an organization consists of designing an organizational memory (Abel et al., 2004).

There have been many researches done on knowledge management itself. Data can be transferred, information can be shared but knowledge is an attribute of people or communities or societies. According to Dougherty (1999), knowledge only exists because of people. Knowledge comes as a person uses information and combines it with their personal experience. Much of the knowledge one acquires and gathers in one's head has its own value, and it is that which makes each of us unique and valuable to the society as a whole and to organizations. Tobias (2000) and Trepper (2000) have also suggested that the two greatest assets that companies have are the people that work with and knowledge in their workers' heads.

Drucker (1993) describes knowledge as the only meaningful resource in a knowledge society. He further stresses that knowledge is not impersonal like money. Knowledge does not reside in a book, data bank, a software programme. They contain only information. Knowledge is always embroiled in a person, taught and learned by a person, used or misused by a person.

McAdam and O'Dell (2000) have undertaken a study on the perception and the use of knowledge management in both public and private sector. They have used Demarest's socially constructed models as their model, as they assume a wide definition of knowledge and represent knowledge as

being intrinsically linked to the social and learning processes within the organization.

Al-Athari and Zairi (2001) have carried out another research project on knowledge management in both private and public sector organizations. Their study examined the actual situation on the availability of knowledge management systems in 77 Kuwait Organizations.

According to Civi (2000), many companies are beginning to understand that the knowledge of their employees is the most valuable asset. Knowledge management has thus far been addressed at either philosophical or technological level, with little pragmatic discussion on how knowledge can be managed and used more effectively on a daily basis.

Liebowitz and Chen (2003) have also conducted another study on knowledge management issues in public sector organizations. They investigated on how knowledge management could build and nurture a knowledge sharing culture in an organization.

Bender and Fish (2000) recognize that knowledge originates in the head of an individual and builds on information that is transformed and enriched by personal experience, beliefs and values with decision and action-relevant meaning. It is information interpreted by the individual and applied to the purpose for which it is needed. The knowledge formed by an individual will differ from another person receiving the same information. Therefore, Bender and Fish (2000) conclude that knowledge is the mental state of ideas, facts, concepts, data and techniques, recorded in an individual's memory. It involves the processing, creation or use of information in the mind of the individual (Kirchner, 1997). Unlike traditional raw material, knowledge usually is not coded, edited, inventoried and stacked in a warehouse for employees to use as needed. It is scattered, messy and easy to lose (Galagan, 1997).

According to several researchers, explicit knowledge is characterized by its ability to be expressed as a word or number, in the form of hard data, scientific formulas, manuals, computer files, documents, patents and standardized procedures or universal starting points that can be easily transferred and spread. On the other hand, implicit knowledge is difficult to formalize and therefore difficult to transfer and spread. It is mainly located in people's hearts and heads. Implicit knowledge is what is in our heads and explicit knowledge is what we have codified.

Nonaka and Takeuchi (1995) have argued that a successful knowledge management programme needs to convert tacit knowledge into explicit knowledge in order to share it and for individuals and groups to internalize and make personally meaningful codified knowledge once it is retrieved from the knowledge management system.

According to Civi (2000), Knowledge originates in human being; a computer cannot create it. The only sustainable advantage of organization is what people know and what they do with it. It is the most important resource a company has that is worth more than land, labour and capital and unlike those traditional assets, knowledge does not diminish in value. It actually represents 75 percent of a company's worth, but does not get a place in the company's balance sheet.

Organization Memory

Organizational Memory defines a comprehensive computer system which captures a company's accumulated know-how and other forms of knowledge assets and makes them available to enhance the efficiency and effectiveness of knowledge-intensive work processes (Vasconcelos, 2000). Furthermore, OM without fail supports the continuous storage manipulation of an organization knowledge (Vasconcelos et al 2002).

Ontology Modelling

Ontology is a model that is populated by concepts and it is organized in a particular hierarchy that represents the theories about real world objects of interest, the relations between them in a certain domain and properties of objects (Vasconcelos, 2000). Protégé 2000 is one of the available ontology which is an open source tool that helps users on the construction of large electronic knowledge bases. This tool enables developers to create and edit domain ontologies.

Among the advantages of Ontology are (1) The ability to share common understanding of the structure of information among people; (2) The ability to reuse domain knowledge and (3) The ability to make the domain explicit.

Outcome Base Education (OBE)

Outcome Based Education (OBE) is an educational process in achieving specified outcomes concerning the students' learning abilities. Both the curricula and education structure are designed in such a way to achieve the capabilities and the qualities that a student should have.

MMU Syllabus contents are modeled and structured from both MMU and MQA guidelines. This structured format will refer as their lecture plan and syllabus. The OBE has since been introduced and implemented by all faculties and departments of Multimedia University (MMU) in the late 2005.

Theoretical Framework

Since knowledge management is an emerging field, there has been no single set of widely recognized and empirically validated criteria for evaluating the successful contributing factors for knowledge management practice. Therefore, in line with the trend toward examining more fully integrated models of the knowledge management contributing factors, a set of variables, taken solely from one perspective, may explain only a small proportion of the variance in how well the factors contribute to knowledge management practice. Moreover, there is little statistical evidence that the proposed factors affect the knowledge management practice, these factors need to be tested especially in the Malaysian context.

The framework is divided into two parts: independent variables and dependent variables. The theoretical framework of this research study consists of a dependent variables (Knowledge management practice among academic in CDP, MMU), and three independent variables (MQA requirements, academic's commitment and University's requirement). Refer to Figure 1.

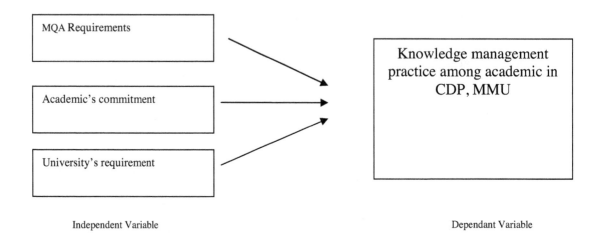

Independent Variable   Dependant Variable

Fig. 1. Theoretical Framework

Dependable variable

Knowledge management practice among academic in CDP, MMU.

Independent variable

a) MQA's requirement

b) Academic's commitment

c) University's requirement

Hypotheses

- **HA1**: MQA's requirement is a contributing factor that encourages knowledge management practice among academics in CDP, MMU.

- **HA2**: Academics' commitment is a contributing factor that encourages knowledge management practice among academics in CDP, MMU.

- **HA3**: University's requirement is a contributing factor that encourages knowledge management practice among academics in CDP, MMU.

Methodology

The population of interest comprises of the academic from CDP, MMU. Simple random sampling technique will be used to select sample. It is an unrestricted probability sampling design whereby every element in the population has known an equal chance of being selected as a subject. A target of 40 academics from CDP, MMU will be randomly selected to be sample for this study.

The time dimension of research would be cross-sectional due to the fact that this study can be carried out in which data are gathered just once in order to meet the research objective.

The data collection for this research will be done through a quantitative nature that is based on survey technique. The survey will be carried out through a self administered and e-mailed questionnaire which is meant to be answered by the academics in CDP, MMU.

The questionnaire consists of five sections in which the academics will be asked to fill up. Section A consists of the individual respondents's demographic characteristics and a nominal scale is used to measure the answers.

Section B is asking the respondents to state their agreement/disagreement on the current issues of knowledge management as adopted from Choi's (2000) study. Each section is cored using a five-point Likert scale. One of the questions in this section is negatively worded because according to Sekaran (2003), instead of phrasing all questions positively, it is advisable to include some negatively worded questions so that the tendency in respondents to mechanically circle the points toward one end of the scale is minimized, especially when the questions are gauging on the respondents subjective feelings such as perception.

Section C is designed to draw information on respondents' perceived importance and the degree of implementation of the practices of knowledge management in their organization. Ernst & Young, Delphi Group and Choi's (2000) study on important factors affecting the implementation of knowledge management in organization, is included in this section. Once again, interval scale is used using a five-point Likert scale.

Section D uses Chois (2000) measurement scales on how knowledge management in general, contributes to organizational competitiveness in Malaysian firms using the five-point Likert scale to seek respondents' opinions.

Finally, Section E describes the potential benefits from implementing knowledge management practices in organization. These items were adopted from Bixley's (2000) study. Interval scale is used where the items in the survey uses five-point Likert scale.

Once the above three hypotheses have been proven, the syllabus management system will be introduced to the academics in CDP, MMU to practice knowledge management.

Syllabus Management System

CDP, Syllabus flow

Currently, CDP is offering 7 different diploma programmes with the duration course of 7 trimesters for each programme. The components of the course structure are divided into 4 parts, which are Mathematics, Core/Major, Electives and University or LAN subjects. Diploma in Information Technology is one of the pioneer courses offered in CDP, offering 28 subjects. Every two trimesters, coordinators are responsible to update the syllabi. The syllabus information is divided into two section; static and variable information. Static information consists of attributes that are fixed and can only be changed upon the approval of MQA. Example of static attributes are subject name and subject code. Variable information consists of attributes that need to be updated from time to time such as reading materials, learning outcomes of the subject, and details of the subject. Besides that, the coordinator needs to update the version and rename the file based on the current version - current month and year. Unfortunately, there are no records of who updates the syllabus. All archived syllabi are then kept by the Manager.

System Flow

Interviews were conducted to 6 people, 4 programme coordinators, one Manager and one Deputy Director of CDP. Requirements and the functionalities of the system are identified and the data collected will represent the information that needs to be captured and represented. UML is used to reflect the system flow. Figure 1 shows the basic flow of the Syllabus Management System, CD.

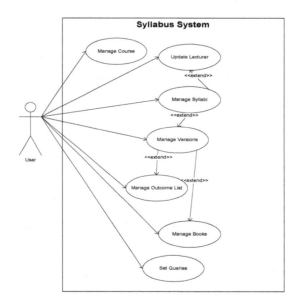

Fig 2. Use Case Diagram of the Syllabus Management System.

Analysis

Step 1. Domain and scope of the ontology

In the syllabus domain, the following are some of the possible competency questions that need to be fulfilled:

• What is/are the subject(s) offered for that trimester?

• What is/are the requisite(s) for the subject?

• What is/are the pre-requisite(s) for the subject?

• Who is/are the lecturer(s) teaching this particular subject?

Judging from the above list of questions, the ontology will include the information on various syllabus characteristics, pre-requisites, requisites, versions, lecturer's information, subject listing, and text/reference books' information.

Step 2. Consider Reusing Existing Ontology

Currently, there is no existing syllabus ontology found or made available through the internet.

Step 3. Enumerate Important Terms in The Ontology

With the help of the guidelines given, some important syllabus terms have been identified. Among the terms are versions, prerequisites, requisites, lecturers, text book, assessment, and many more.

Design

Step 1: Defining the Class Hierarch

In the next stage, classes are created and are arranged in a taxonomical manner as shown in figure 2. Figure 3, represents the object-relationship diagram representing the Syllabus Management System.

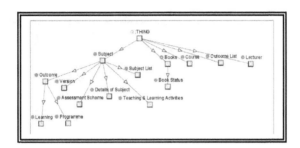

Fig 3. Taxonomy Classes for the Syllabus Management System.

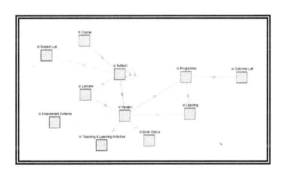

Fig 4. An Object-Relationship Diagram for the Syllabus Management System, CDP

Step 2: Defining Facets and Properties of Slots

The slots are created, identified and are made available in each class. These terms cover the information needed to be captured in the classes defined earlier which includes, a subject code, subject name, version id, credit hours, lecturer, and many more. There are slots that carry different facets describing the values it describes, the type, allowed values, cardinality and other features that it can support. These slots consist of intrinsic properties, extrinsic properties, parts of the object structured and as a relationship to other instances and slots.

Implementation

In the implementation stage, prototypes are designed and instances are created by filling in all the information. Once all this is done, queries are generated to determine the results that enable to answer the competency questions that have been set in the analysis phase.

Step 1: Prototype and Instances.

Layout of the subject is constructed by using the default Protégé 2000 graphical user interface (GUI).

Step 2: Creating Instances and Filling in Information.

Once the prototypes have been developed, instances are created and filled in desired input information.

Testing

The system prototype and the system functionalities are presented to the six committees who determine whether the system is working accordingly. Queries are created to show the results produced. The results showed that all the basic requirements have been fulfilled.

System Limitation

The limitation of the Syllabus Management System are as follows: (1) The system does not focus on the security expect as this is the starting phase of the project development; (2) The system only caters for the syllabi offered in Diploma of Information of Technology (DIT). Therefore, the course offered is based on the DIT course structure; (3) The buttons of add, edit, delete and view button are provided by the Protégé GUI, therefore, it is difficult for the users to understand the icon functionalities; (4) the system does not include the details of topics taught in lectures, tutorial and laboratories, and the hours spend on the defined topics; (5) The system could not display the whole course structure of DIT and (6) There are no access level implemented in this system.

Future Enhancement

The system can be improved if the following is implemented: (1) The system should capture more information of the syllabus such as the details of the topics taught and the hours spend; (2) The system should be able to show the whole course structure; (3) The system should implement the access level and (5) The system should be able to expand the domain so that it will be able to support more competency questions.

Significance/Contributions

The implication of this study is significant because it focuses directly towards the contributing factors that encourage knowledge management practice among academics in CDP, MMU. Hence, it allows knowledge management researchers to gauge the current state of knowledge management research in a systematic and practical manner. In addition, the results of this study will be able to provide an insight into what are the overall perception of knowledge management and how various factors affect the successful implementation of knowledge management practice and organizational performance among academics in CDP, MMU.

More importantly, knowledge management makes the transition from concept to practice; attention must turn to the ways in which academic practitioners can implement the growing body of theory. The findings of this study contribute to the practice of knowledge management among the academics in CDP, MMU, whereby this research provides an opportunity to the practitioners to undergo a self-check for the various important knowledge management areas that this research intend to study. Moreover, once the hypothesis has been proven through the research, a syllabus management system is used to practice knowledge sharing and this system is a significant tool which will iteratively and gradually improve and support for the entire programmes offered in CDP, MMU, Malacca. Changes are easily made to suit both the CDP and MQA requirements.

Among the objective of this project is to use the ontology in structuring the Syllabus Management System for CDP. Nevertheless, ontology allows the flexibility and reusability of domain knowledge that makes it possible to change the assumptions if the domain changes. There is no one correct way or a static method in modeling a domain. There are several alternatives to choose from but the best solutions have been reflected in the system requirements.

The model created is flexible, allowing the integration between ontology and application and the ability to extend the class hierarchy without restricting its depth or breadth (A.Abu-Hanna et al, 2005). The syllabus management system is flexible and adaptable and it can be easily suited and integrated to the proven hypothesis for knowledge management practice and sharing of knowledge among academics in CDP, MMU.

Conclusion

Overall, the main objective of this study is to identify and prove that MQA's requirements, Academics' commitment and University's requirement are among the main contributing factors for knowledge management practice among academic in CDP, MMU. A Syllabus management system helps in creating and managing the knowledge management practice and sharing knowledge among the academics. The Syllabus management System is developed using Protégé 2000. However, the system only caters for the Diploma Information Technology programme and only certain information is captured to represent the system requirements. Ontology development is one iterative process that permits knowledge reusability and is the better engineering of knowledge based system with respect to acquisition, verification and maintenance.

This research will be the beginning for the development of the Syllabus Management System for the CDP, MMU, Malacca. Using ontology as guidance in structuring the system, allows the opportunities to expand and reuse the system for other programmes offered in CDP. Besides that, with the assistance of ontology tools, it helps the system to manage, distribute, capture and represent the knowledge base of the syllabi and the future Outcome Based Learning activities.

References

A.Abu-Hanna, R. Cornet, Nicolette K., Monica C. & S. Tu (2005). 'PROTE´ GE´ as a Vehiclef Developing Medical Terminological System,' Department of Medical Informatics, AMC-University of Amsterdam.

Abel, M., Benayache, A., Lenne, D., Moulin, C., Barry, C., & Chaput, B. (2004), "Ontology-based Organizational Memory for E-learning," *Educational Technology & Society*, Vol. 7 (4), pp.98-111.

Al-Athari, A. & Zairi,M. (2001). "Building Benchmarking Competence through Knowledge Management Capability: An Empirical Study of the Kuwaiti Context," *Benchmarking: An International Journal*, 4 (3), 70-80.

Bender,S. & Fish, A. (2000). "The Transfer of Knowledge and the Retention of Expertise: The Continuing Need for Global Assignment," *Journal of Knowledge Management*, 4 (2), 125-137.

Bixler, C. H. (2000). 'Creating a Dynamic Knowledge Management Maturity Continuum for Increased Enterprise Performance and Innovation,' Unpublished academic dissertation, The George Washington Unversity.

Choi, Y. S. (2000). "An Empirical Study of Factors Affecting Successful Implementation of Knowledge Management," Unpublished academic dissertation, University Nebraska.

Chong, S. C. (2005). 'Implementing of Knowledge Management among Malaysian ICT Companies: An Empirical Study of Success Factors and Organizational Performance,' Unpublished doctoral dissertation, Multimedia University, Malaysia.

Civi, E. (2000). 'Knowledge Management as a Competitive Asset: A Review Marketing Intelligence and Planning,' *Journal of Knowledge Management*, 18 (4), 166-174.

Dougherty, V. (1999). "Knowledge is about People not Databases," *Industrial Commercial Training*, 31 (7), 262-266.

Drucker, P. (1993). 'The Post Capitalist Society,' New York: *Harper Business Press*.

Galagan, P. (1997). "Smart Companies," *Training and Development*, 51 (12), 20-26.

Kirchner, D. (1997). 'The Shape of Things to Come,' *World Traveler*, May, 54-77.

Liebowitz, J. & Chen, Y. (2003). "Knowledge Sharing Proficiencies: The Key to Knowledge Management," *Handbook of Knowledge Management 1: Knowledge Matter, Springer-Verlag*, 409-424.

McAdam, R. & O'Dell, C. (2000). "A Comparison of Public and Private Sector Perceptions and Use of Knowledge Management," *Journal of European Training*, 24 (6), 317-329.

Nonaka, I., & Takeuchi, H. (1995). 'The Knowledge Creating Company: How Japanese Companies Create Dynamics of Innovation,' New York: Oxford University Press.

Sekaran, U. (2003). 'Research Methods for Business: A Skill-Building Approach,' *New York: John-Wiley and Sons*.

Steffen, S., Rudi, S. S., Schnurr, H. P. & Sure, Y., (January/February 2001), "Knowledge Processes and Ontologies," *Knowledge Management*, pp.26- 34.

Thorsten, L. & Sure, Y. (2002). "Introducing Ontology-based Skills Management at a large Insurance Company,"

Tobiaz, Z. (2000). 'Champions ok knowledge,' *Computer World*, 34 (40), p.84.

15

Knowledge Management Innovation: Perspectives from the Islamic Development Bank

Amir Raslan Abu Bakar[1] and Rugayah Hashim[2]

[1]Islamic Development Bank, Jeddah, Saudi Arabia

[2]Universiti Teknlologi Mara, Shah Alam, Selangor, Malaysia

Abstract

International financial institutions should no longer rely on traditional ways to conduct business. By innovating on legacy platforms, organizations are able to be on the competitive edge. Information is power and managing information or knowledge will ensure profits are maximized and competitive business advantage is attained. In the case of the Islamic Development Bank (IDB), becoming the leader in the international banking industry requires smart leveraging of innovation in knowledge management. By exploring these innovative capabilities, organization information processes will enhance the routine activities for IDB while simultaneously positioning itself strategically in the eyes of the world as the leading Islamic financial institution that serves the interests of the people to its fullest potential. Consequently, key performance indicators (KPI) can then be identified and applied to ensure objectives are achieved. The outcome of this paper would be of interests not only to the IDB's top management but also to academics, advocates of KM and innovation and would ultimately contribute towards breadth of knowledge within these two areas for further in-depth, empirical studies

Keywords: innovation, knowledge management, competitive advantage

Introduction

On August 5, 2011, Elliot (2011) reported that the most serious global economic crisis has hit the lowest grade since the Great Depression. Again, the crisis was triggered by the newly downgraded US debt status (Elliot, 2011). Previously in May 2009, the world economy was also in a deep recession, which started with the financial crisis in the United States (Boston Consulting Group, 2009; Elliot, 2011). This crisis occurred through a combination of low interest rate policy, deregulation of non-bank financial institutions and the massive growth of the unregulated derivatives markets (Boston Consulting Group, 2009). During these tumultuous times, survival of financial companies becomes critical. Organizations across the world including IDB have to be more innovative either in cutting cost of operations, reengineering the production processes and inventing new products or services to stay relevant in the market. Knowledge has been identified as an important element which is critical for organizations to be innovative (Ho, 2007).

Nevertheless, the challenge for organizations to stay competitive is becoming more and more difficult these days due to the continuous global economic crisis. Adopting and executing the right strategy is becoming more imminent (Morgan, Levitt and Malek, 2007). Furthermore, organizations are forced to think 'outside of the box' in order to position themselves strategically in the market. The challenge for organizations to stay competitive is becoming more and more difficult these days due to the current economic crisis. Adopting and executing the right strategy is becoming more imminent (Morgan, Levitt and Malek, 2007). The tolerance and margin for error are less compromising as small failures have bigger business implications which would affect bottom lines and force organizations to be out of business (Andrew and Sirkin, 2006). To prepare for the inevitable battle, organizations are leveraging on the knowledge that they inherit and acquire from outside to be as innovative as possible in providing solutions. Knowledge and innovation are key intangible assets which are priceless in this world of borderless economy (Uhlaner, van Stel, Maijaard and Folkeringa, 2007).

As awareness of capturing tacit and explicit knowledge is becoming increasingly essential among business communities these days, the need to conceptualize the scope, properties and remit of knowledge management (KM) has become apparent. Strategically, organizations are adopting knowledge management and innovation strategy as key ingredients to re-align their business directions, identifying ideas, managing risks efficiently, monitoring and managing activities effectively (Dhondt, 2003).

The theory of economic growth postulates that innovation is a primary source of an organization productivity growth and cost cuttings. Innovation is the outcome of organizations' efforts to produce new or improved products, introduce more efficient productive processes and implement organizational or managerial changes or new marketing and design processes (Dhondt, 2003). Innovation stands out as one excellent

objective of management activity in general, and knowledge management specifically (Firestone, 2001).

Andrew and Sirkin, (2006) suggest that the challenge of innovation is not in the lack of ideas but rather in successfully managing the innovation so that it delivers the required return on the organizations' money, time and people. This study reveals how knowledge management and innovation form a strategic combination to lay the foundation for organizations to achieve competitive advantage.

Significance of Study

The Islamic Development Bank (IDB) is undergoing a new transformation agenda in conjunction with its new Vision 1440H. There are nine strategic key thrusts mandated under this vision. Broadly guided by the end goals in mind, IDB would be more stringent in its endeavor to alleviate the standard of living and prosper the Ummah across the world through Islamic finance, syariah compliance initiatives and economic integration that mobilizes resources between member countries and Islamic communities.

Background and Literature Review

These days, many reports have been published to inform that the global business landscape is littered with expensive, well intended strategies that failed due to many reasons such as poor content management, lack of coordination, no buy-in and poor execution. Any business entity's strategy describes how it intends to create value for its shareholders, customers and citizens as are the vision and mission of IDB.

About IDB, it was established in 1973 to "foster the economic development and social progress of member countries and Muslim communities individually as well as jointly in accordance with the principles of Shari'ah, that is, Islamic Law." (Islamic Development Bank, 2011).

Scholars in management are suggesting that appreciating the importance of knowledge and applying it in the right manner is becoming more and more inevitable these days for organisations to be innovative in their business processes and products or service offerings. The application of balance scorecard (BSC) which was introduced by Kaplan and Norton (1996, 2004) offers a framework for translating strategies into value creation which can be monitored effectively through:

- translating the vision into operational goals;

- communicating the vision and linking it to individual performance;

- business planning;

- feedback and learning, and adjusting the strategy accordingly.

Nevertheless, within the scope of IDB, the paper, albeit briefly, intends to explore:

- the meaning of knowledge and its importance;

- the concept of KM;

- how organizations can leverage on knowledge

- how can the application of KM assist organizations to position themselves competitively

- the meaning of innovation and its importance

- how KM can lead companies to be innovative

- how can the concept of Stage-Gate be applied to innovation process

- the critical success factors in implementing KM and innovation

Organizations are forced to think 'outside of the box' in order to position themselves strategically in the market. The tolerance and margin for error are less compromising as small failures have bigger business implications which not only would be effecting bottom lines but also force organizations to be out of business (Andrew and Sirkin, 2006). To prepare for the inevitable battle, organizations are leveraging on the knowledge that they inherit and acquire from outside to be as innovative as possible in their solutions providing. Knowledge and innovation are key intangible assets which could be priceless in this world of borderless economy (Uhlaner, van Stel, Maijaard and Folkeringa, 2007). As awareness of capturing tacit and explicit knowledge is becoming increasingly essential among business communities these days, the need to conceptualize the scope, properties and remit of knowledge management (KM) has become apparent. Strategically, organizations are adopting knowledge management and innovation strategy as key ingredients to re-align their business directions, identifying ideas, managing risks efficiently, monitoring and managing activities effectively (Dhondt, 2003). The theory of economic growth postulates that innovation is a primary source of an organization productivity growth and cost cuttings. Innovation is the outcome of organizations' efforts to produce new or improved products, introduce more efficient productive processes and implement organizational or managerial changes or new marketing and design processes (Dhondt, 2003). Innovation stands out as one excellent objective of management activity in general, and knowledge management specifically (Firestone, 2001). Andrew and Sirkin, (2006) suggest that the challenge of innovation is not in the lack of ideas but rather in successfully managing the innovation so that it delivers the required return on the organizations' money, time and people. This study reveals how knowledge management and innovation form a strategic combination to lay the foundation for organizations to achieve competitive advantage.

The society we are in these days is gradually transforming from industry-based to knowledge-intensive (Van de Ven, 2004). The transformation is coerced by two emerging forces: the changing global economy interactions and inexorable technological enhancement (Chan, Deng, Peng and Xi, 2006). The accelerating intensification of IT assisted by wide usage of internet has developed the business world into borderless economy (Lu, Yuan, Tsang and Peng, 2008).

The world is now experiencing a radical transformation from a mass production system where the principal source of value was human labor and now to the epoch of 'innovation-mediated production' where the principal component of value creation, productivity and economic growth is knowledge (Andrew and Sirkin, 2006). More and more talented people are now being hired globally with the hope that they would be able to make full use of the knowledge that the organizations have and be innovative in their business processes and solution offerings.

Shedroff (2001) suggests that the greater an organization understands a particular subject that it is focusing on, the more it would be able to weave past experiences into new knowledge by absorbing, doing, interacting, and reflecting. Observing the development in management studies, there is a clear transformation process from the state of acknowledging the importance of data to where the world is now moving, that is, having good knowledge alone is not sufficient. Knowledge must be processed further so that it can be applied successfully in order to meet the objectives of the organizations.

Knowledge is so crucial that it has become a key endeavor towards achieving competitive advantage (Scarbrough, 2003). The problem that most organizations are facing these days is to identify and capture the relevant knowledge and then apply it in the right direction (wisdom) so that they can be innovative in their processes or output which would have substantial bearing towards reduction in cost of production or capitalizing on their niches. Also, in order to manage effectively, performance of input must be able to be measured efficiently (Fombrun and van Riel, 2004).

KM plays a big role in innovation process (Tiwana, 2003), thus, this section will analyze some of the major roles that KM would be able to facilitate and provide better grounds for innovation process to be more successful for the Islamic Development Bank (IDB). As posited by Cardinal et al (2001), tacit knowledge is converted to explicit knowledge, thus, KM provides the platform as well as the processes to ensure that tacit knowledge becomes explicit knowledge, for example, the codification platforms for discussion databases or online collaborative communities of practice. Within the realms of IDB, the capturing of tacit knowledge can be done during the sharing events such as breakfast briefings which can be converted to electronic form where the knowledge can be organized and retrieved for later use. This adds a lot of value to the organization as it is disseminated to the staff what knowledge is available, and it is retrievable for future re-use. Tacit knowledge sharing is critical for IDB's innovation capability but the replication of knowledge-based competitive advantage is inhibited by two factors:

- causal ambiguity leads to specific practices or inputs for replication being unknown; and

- Social complexity or unique organization history that produces the knowledge makes it difficult to replicate.

Even though explicit knowledge is not as dominant as tacit knowledge, it is still considered to be an important component of innovation. In developed science processes, explicit knowledge features quite strongly in the research and development (R&D) processes as there is a rich exchange of tacit knowledge taking place, which IDB should heed to. Moreover, KM provides the tools, processes and platforms to ensure knowledge availability and accessibility, for example through structuring of the knowledge base. KM can also ensure that explicit knowledge, which can be used as an input to the innovation process, is gathered internally and externally. Finally, KM also provides the means of ensuring the leverage of knowledge and to determine the gaps in the explicit knowledge base that IDB could potentially impact IDB's innovation programs.

Furthermore, in managing innovation, KM plays an important role that enables collaboration. Collaboration requires suppliers, customers, and employees to form knowledge sharing communities within and across organizational boundaries to achieve a shared business objective for communal benefits. In addition, internal and external collaboration plays an important role in the transfer of tacit knowledge and building collective know-how (Pyka, 2002) through online collaboration forums such as intranets and extranets which are readily available at IDB. These collaboration forums are extremely valuable because they ensure the codification of knowledge utilized as input to the innovation process. The stronger the relationship between collaboration partners, the greater the extent of the tacit knowledge transfer (Scarbrough, 2003). Gathering tacit knowledge from collaboration partners could potentially reduce risk and cost in innovation by ensuring a first-time-right approach, thus shortening development cycles and ensuring effective innovation.

Needless to say, time is of essence for all business entities. Through knowledge integration, timely insights can be made available to be drawn at the right juncture for sense making, that is, knowledge can be exchanged, shared, evolved, refined and made available at the point of need. With that in mind, without accurate information and KM to underpin knowledge integration, IDB is at risk in respect of inefficiently utilizing knowledge as resource for innovation.

Doubtlessly, KM provides an environment for knowledge creation, sharing and collaboration. Gloet and Terziovski (2004) concluded that there is a significant and positive relationship between KM practices and innovation performance, and those organizations therefore, should strive for an integrated approach towards KM, which assists in building a corporate culture, in order to maximize innovation performance leading to competitive advantage. To re-emphasize, KM ensures the availability and accessibility of both tacit and explicit knowledge used in the innovation process using knowledge organization and retrieval skills and tools, such as taxonomies. It allows the organization to retrieve knowledge in a structured way according to the unique structures and value chain of the organization. It also provides search facilities and tools to enable IDB staff to search for the knowledge required in the innovation process.

In KM initiatives, platforms, tools and processes to ensure integration of IDB's knowledge base can be established. Through KM structures such as taxonomies, KM can ensure the integration of the corporate knowledge base. As a result, the staff would have an integrated view of what knowledge is available, where it can be accessed, and gaps in the knowledge base. This is extremely important in the innovation process for two reasons:

- to ensure that knowledge as a resource is utilized to its maximum benefit; and

- at the same time to ensure that knowledge is not recreated in the innovation process.

Due to the knowledge-driven culture which is embedded in KM, IDB's innovations can be incubated. Knowledge sharing is enhanced by a culture where the role of knowledge, KM, innovation and creative thinking is encouraged. KM programs usually have a strong knowledge culture element through which IDB's organizational culture of knowledge generation and sharing can be given greater attention. Because of the culture within KM and innovation, creativity and learning through mistakes are encouraged and valued.

Furthermore, IDB employees are able to increase their skill levels and knowledge both formally and informally through knowledge accessibility, sharing and a smooth workflow. Increasing the staff's skills would eventually provide better chances for IDB to have quality innovations. The flow of knowledge across functional boundaries ensures that a wider base of knowledge is available to the staff than only the knowledge they use in their day-to-day activities. Therefore, IDB's staff will have a wider frame of reference of the context in which they work and will therefore be able to innovate more efficiently.

KM assists in identifying and understanding the organizational context, that is, it provides organizational context to the body of knowledge in the organization. Every organization has its own unique corporate memory including IDB. The structures provided to organize and retrieve knowledge from the corporate knowledge base will provide a unique context to each particular organizational knowledge base. Provision of organizational context is critical in the innovation process, as innovation in the organization also takes place within a very specific business context (Tidd and Bessant, 2009).

Lastly, KM plays an important role in identifying gaps in the knowledge base and provides processes to fill the gaps in order to aid innovation. Through the structured provision of access to knowledge, KM provides an overview of what is available in IDB. As a result, this allows IDB's management to understand which knowledge area is lacking and to systematically build the knowledge base in these areas.

Discussion and Conclusion

With IDB in mind, it is important to reiterate that KM systems alone do not possess the qualities required to provide organizations with sustainable competitive advantage but the bundling of KM systems with other organization resources and core competencies is the key to develop and maintain sustainable competitive advantage through product and process innovation. In such a position, KM systems play a major role in the conversion of learning capabilities and core competencies into sustainable advantage by enabling and revitalizing organizational learning and resource development processes, which are:

- to create, build and maintain competitive advantage through utilization of knowledge and through collaboration practices. KM can facilitate such collaboration. Provided that there is a close collaborative relationships between organizations, the application of KM can be applied as a cross cutting tool that intrigues across organizational boundaries to leverage on the knowledge that they have and provide shared sustained innovation and competitive advantage;

- to reduce complexity in the innovation process, and manage knowledge as a resource will consequently be of significant importance. Innovation is extremely dependent on the availability of knowledge and therefore the complexity created by the explosion of richness and reach of knowledge has to be recognized and managed; and

- to integrate internal and external knowledge which the organization can grasp. This entails that timely insights can be made available at the appropriate juncture so that knowledge can be exchanged, shared, evolved, refined and made available at the point of need.

Knowledge integration via KM platforms, tools and processes must therefore facilitate reflection and dialogue to allow personal and organizational learning and innovation.

Impact studies in this area may be extremely valuable, especially to IDB's distinct knowledge management and innovation programs. It is important for both innovation and knowledge management professionals in IDB to understand the systemic relationship between these concepts and the value that it can generate in respect of creating and maintaining sustainable competitive advantage for IDB and other financial institutions. Throughout this paper, benefits of innovation are clearly discussed, hence, it is important to note that, there are also some crucial indirect benefits which IDB should capitalize from the innovation processes such as:

- *branding*: innovation can enhance the brand – IDB's famous presence and acronym is a brand in itself, thereby attracting more customers and enabling IDB to charge relevant premium for their products and services;

- *ecosystem*: IDB's innovators can create exceptionally strong ecosystems of partners and associated organizations, enabling them to leverage their position strategically;

- *knowledge*: the innovation process always produces knowledge, some of which can usually be put to work in more than one way; and

- *Organization*: being innovative allows IDB to attract and retain more of the best people, or at least more of the most innovative ones.

In conclusion, the theory of economic growth postulates that innovation is a primary source of an organization's productivity growth and cost cuttings. As discussed throughout this paper, innovation is the outcome of an organization's efforts, which in IDB's case is to produce new or improved services, introduce more efficient processes and implement organizational or managerial changes or new marketing and design processes. The prime reason for companies to focus on knowledge management is that knowledge is regarded as the driving force for the organizations of the future. This study has proven that it makes good sense that knowledge management and innovation would from a strategic combination for IDB to achieve competitive advantage where the challenge of innovation is not because of lack of ideas, but rather of successfully managing the innovation so that it delivers the required return on the organizations' money, time and people Andrew and Sirkin, (2006). This paper has looked into how tacit and implicit knowledge in an organization are captured, and the idea creation process, which is the key element in innovation, can be effectively filtered and managed by using a best practice method called Stage-Gate method. Capitalizing from the analyses of this paper, further research could be conducted to see how the Balance Scorecard (BSC) and key performance indicators (KPI) can be applied into innovation processes. A holistic action plan is perhaps established to encompass knowledge management, innovation, BSC and KPI into an overall strategy map for IDB.

References

Andrew, J. P. & Sirkin, H. L. (2006). "Payback: Reaping the Rewards of Innovation," *Harvard Business School Press*.

Boston Consulting Group (2009). Accessed December 20, 2004 at http://www.bcg.com/

Cardinal, L. B., Allessandri, T. M. & Turner, S. F. (2001). "Knowledge Codifiability, Resources, and Science Based Innovation," *Journal of Knowledge Management*, 5 (2). pp.195-204.

Chan, N. H., Deng, S.-J., Peng, L. & Xi, Z. (2007). "Interval Estimation for the Conditional Value-At-Risk Based on GARCH Models with

Heavy Tailed Innovations," *Journal of Econometrics* 137(2). 556 - 576.

Clark, D. (2004). Accessed December 20, 2004   at http://www.nwlink.com/~donclark/performance/understanding.html

Dhondt, S. (2003). "Knowledge Management, Innovation and Creativity," *TNO-report*.

Elliot, L. (2011). "Global Financial Crisis: Five Key Stages 2007-2011," Retrieved August 15, 2011   at http://www.guardian.co.uk/business/2011/aug/07/global-financial-crisis-key-stages/print

Firestone, J. M. (2001). "Key Issues in Knowledge Management," *Journal of the KMCI*.

Gloet, M. & Terziovski, M. (2004). "Exploring the Relationship Between Km Practices and Innovation Performance," *Journal of Manufacturing Technology Management*, Vol. 15 No.5, pp.402-9.

Gottschalk, P. (2007). "Knowledge Management Systems – Value Shop Creation," *Idea Group Publishing*.

Ho, D. (2007). "Research, Innovation and Knowledge Management: The ICT Factor," *UNESCO*.

Islamic Development Bank (IDB). (2011). Accessed   at http://www.isdb.org/irj/portal/anonymous?NavigationTarget=navurl://24de0d5f10da906da85e96ac356b7af0

Kaplan, R. S. & Norton D. P. (1996). 'Balance Scorecard,' *Harvard Business School Press*.

Kaplan, R. S. & Norton D. P. (2004). "Strategy Maps," *Harvard Business School Press*.

Lu, Y., Tsang, E. W. K. & Peng, M. W. (2008). "Knowledge Management and Innovation Strategy in the Asia Pacific: Toward an

Institution-Based View," *Asia Pacific Journal of Management*, 25 (3): 361-374.

Morgan M., Levitt, R. E. & Malek, W. (2007). "Executing Your Strategy," *Harvard Business School Press*.

Pyka, A. (2002). "Innovation Networks in Economics: From the Incentive-Based to the Knowledge Based Approaches," *European Journal of Innovation Management*, Vol. 5 No.3, pp.152-63.

Scarbrough, H. (2003). "Knowledge Management, HRM and the Innovation Process," *International Journal of Manpower*, Vol. 24 No.5, pp.501-16.

Shedroff, N. (2001). "Experience Design," *New Riders Publishing*.

Tidd, J. & Bessant, J. (2009). "Managing Innovation: Integrating Technological, Market and Organizational Change," West Sussex, England: *Wiley*.

Tiwana, A. (2003). 'The Knowledge Management Toolkit,' *Prentice Hall*, PTR

Uhlaner, L. M., van Stel, A. J., Meijaard, J. & Folkeringa, M. (2007). "The Relationship between Knowledge Management, Innovation and Firm Performance: Evidence from Dutch SMEs," *EIM Research Report*.

Van de Ven, A. H. & Engleman, R. M. (2004). "Event- and Outcome-Driven Explanations of Entrepreneurship," *Journal of Business Venturing*, Vol. 19, pp. 343-58

16

Towards an Information Intelligence and knowledge Management Process in the Context of Information Technologies

Boulesnane Sabrina, Bouzidi Laïd and Marini Jean-Luc

Jean Moulin - Lyon 3 University, Lyon. France

Abstract

In this paper, a model for information intelligence and knowledge management, in the context of audit and advisory companies in the field of information technologies, is proposed. This model is founded on the exploitation, structuring, analysis and optimisation of strategic information in organisation. The implementation of information intelligence process indicates all the competence, devices and necessary initiatives for allowing information technology professionals to manage customers' needs. The issue of our study is to provide an Assistance System of needs' interpretation. The model herein proposed will influence future research and managerial practices.

Keywords: Information intelligence, Knowledge management, Information technologies, Needs' interpretation.

Introduction

Modern organisations have to permanently evaluate their performance, global coherence and measure their information system efficiency. Organisations lean on audit and advisory companies, as well as on experts in information systems and information technology integration. These companies work in a complex and dynamic socio-economic environment, where visibility is an essential parameter for survival (Laudon and Laudon, 2011; Harris et al, 2008).

Generally, the problem these professionals encounter in Small and Medium-sized Enterprises (SMEs) is needs' interpretation. This problem was identified in a pragmatic study realised within two firms of auditors and advisors, situated in the Rhône-Alpes region of France. In this study, we observe that the heterogeneity between formulated needs (customers' companies) and their identification and interpretation (information technology professionals) generate situations of confusion.

Our interest concerns the analysis of documents provided by consulting in this context. We draw particular attention to the exploitation of a technological lexicon collected in this pragmatic study, in order to identify the semantic incoherences and ambiguities.

The aim of this study is to provide an 'Assistance System of needs' interpretation. The implementation on the ground of this system enables the identification of the research perspectives.

Literature Review

The socio-economic organisations environment is more and more complex, unpredictable and in perpetual transformation. In this unstable and uncertain context, organisations turn to the implementation of information intelligence and knowledge management processes. The focus will be on these processes in the context of audit and advisory companies in information technologies.

Information Intelligence in Organisations

We found in the litterature a variation of definitions of the information intelligence process in organisations (Bellon, 2002; Crowne, 2009). Information intelligence is considered to be the capacity to resolve problems, as a result of strategic knowledge, oriented to the action and the decision. This process requires the presence of relevant and necessary information, to know how to exploit it for problem resolution (Hauch et al, 2005; Poirier, 2000). This approach of production and appropriation of information content allows the optimisation of management processes. The process of

information intelligence leans on several complementary competences (Boisvert, 2010).

The first level is the collection and structuring of the information necessary for the resolution of the problems that the organisations encounter.

The second level consists of analysing the collected information and identifying the strategies of exploitation and information-seeking.

The third level is the management and treatment of information, aimed at creating profitable and exploitable knowledge.

Finally, information intelligence leans on the communication and collaboration between the mobilised actors, in respect of ethical guidelines.

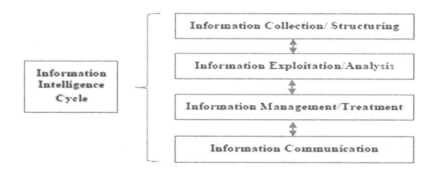

Fig 1. Necessary competence for the Information Intelligence Process

More than ever, the human, organisational, intellectual and information capital exploitations create a superior value in organisations' management. The knowledge generated by strategic and 'intelligent' information exploitation accentuates the performance of 'modern' organisations (Ermine, 2003).

Knowledge Management in Organisations

The information intelligence optimises the knowledge management by the implementation of devices and methods of knowledge valuation. There are many

possible ways to describe and define knowledge management activities. In general, knowledge management is a process consisting of several steps that provide added value in organizations' management practices. According to Šikýř (2010), 'the purpose of knowledge is to improve the storing, creation, sharing and use of knowledge in the organization, and thus improve performance of individual employees and the organization as a whole entity'. (Nanoka, 1994) consider that Knowledge Management is 'context-specific'. The organisation no longer has to limit itself to a role of information consumption; thus guaranteeing its survival and economic development. This concerns all organizations, whatever their size or business sector. Among the professional practices, we are interested in the technologic audit and advisory companies.

Information Intelligence and Knowledge Management in a Context of Audit and Advisory in Information System and Technologies

Firms of audit and advisory accompany organisations in the information systems and technology integration in their functional processes. They bring this help with the condition of assimilating and understanding the customers' needs. It has to be admitted that, often, these conditions are far from being satisfied. The problems of needs' understanding create the suggestion of inadequate solutions to the real needs of the SME.

Within SMEs, it is the 'functional actors', accompanied by technological professionals, who identify the needs. These needs must be the most explicit possible, otherwise the actors risk using incoherent and incomplete information.

Audit and advisory companies intervene in different communities, characterized by a specific language. These professionals have to become acquainted with the customers' jargon. The implementation of information intelligence process in these firms allows them to create a context favourable to a common language use.

Research Context

We realize a study on the ground, which we refined between 2005 and 2011. This study concerns two audit and advisory firms, situated in the Rhône-Alpes region (France), that specialize in the integration of information technologies in SMEs. The problems of needs' comprehension between customers' organisation (needs formulation) and the audit and advisory companies (needs' interpretation) are analysed.

This research work is based on a qualitative approach, leaning on 'participant observation' (DeWalt et al, 1998; Silverman, 2009), and enriched by constant exchanges with professionals of new technologies. The exploration on the ground allowed to analyse this research problem. The implementation of 'focus groups' consisted of consultants and customer organizations, and aimed to obtain information in order to clarify the problematic and identify the research hypothesis.

The present research focuses on an empirical study, based on the analysis of about fifty documents supplied by the studied consulting firms. The analysis is concerned with the technological vocabulary exploited in this pragmatic context. We extracted 100 ambiguous terms and expressions used by 'non-specialist' customers.

We observed that the use of the technological vocabulary was inconstant and the information detected was generally ambiguous. On the one hand, the actors who formulated the needs leaned on different referential terms (customers' organization). Generally, they use the same concept to identify several expressions, characterised by various meanings, in a context of needs' formulation. On the other hand, the audit and advisory companies encounter difficulties in term of needs' interpretation. This situation generates problems of needs' comprehension and exact identification.

The extracted technological vocabulary is the object of a linguistic analysis allowing us to determine the problems of semantic ambiguity. An 'Assistance System of needs' interpretation is proposed, based on the exploitation of a process of information intelligence and knowledge management. This system allows the establishment of a common language in a complex and dynamic activity. The research context is represented in the figure bellow.

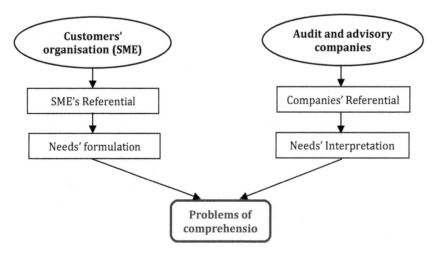

Fig 2. Research Context

Research Hypothesis and Questions

The information produced and exploited in the context of audit and advisory companies, and used for needs' formulation, are complex and dynamic. This is the case owing to several criteria which structure this research hypothesis.

Hypothesis 1: The needs' expression in organisations is carried out by the CIO (Chief Information Officer), if possible. Owing to financial difficulties, SMEs delegate the expression of needs to actors who are not necessarily experts in the technological field (such as, the financial officer, accountant etc.).

Hypothesis 2: In information system, projects involve generally heterogeneous actors (customers and technological actors). The variety and diversity of their referential produce phenomena of 'complexity and semantic heterogeneousness'.

Hypothesis 3: Information technology field evolution creates a rapid transformation of the technological language. This evolution generates confusion in the use of the information technology terminology.

The study allowed analysing the semantic incoherence which exists between the formulation of needs and their interpretation. The constitution of a 'common referential' will help audit and advisory companies to face the problems of needs' interpretation.

Research Design and Methodology

The process of information intelligence leans on complementary phases: collection, exploitation, management and communication of information. All these phases, declined in the present research context, result in the implementation of an Assistance System of needs' interpretation. The system implementation ensures a social cohesion between heterogeneous actors (Boulesnane and Bouzidi, 2009).

Collection and Structuring of Information

We can sum up the first analysis phase by making a list of 100 terms used by 'non-specialist' customers. This corpus was extracted from 50 documents provided by the audit and advisory companies studied. The expressions and terms analysis reveals problems of incoherence and semantic ambiguity.

We cite some words and expression which constitute the lexicon: Database (DB); Database Management Systems; Datamining; Datawerhouse; Decision Support Systems; ERP (Enterprise Resource Planning); Expert System; Information and Communication Technologies (ICT); Informatics system; Information System (IS); Networks; Software; Web technologies etc.

Exploitation and Information-Seeking

In the second phase, a semantic analysis identified some relationships which we consider as 'contextual relations'. Even if the approach is based on language analysis, we were interested more in the pragmatic level. Indeed, we observed that human actors are confused by the concepts, and tend, generally, to associate some terms with a second interpretation level, which we call 'contextual synonymy'.

The treatment and filtering of this corpus allowed collecting a group of terms which are characterized by identical semantic relations. In order to organise and structure our corpus and the relationships between concepts, we use the graph theory. The objective is to give a synthetic overview of the different terms and the contextual relations which connect them. The exploitation of the graph leans on an approach offering two alternatives. The lexicon can be exploited by having either the lexicon (Lexical Database) or the profile of customers (Profile Database) as the point of departure.

By using different alternatives, we were able to exploit the terms extracted from the documents provided by the studied audit and advisory companies.

The proposed approach is consolidated by using tools and information management strategies.

Management and Treatment of Information

The various confusion cases are composed of a set of 'terms candidates', which constitute what we call an 'interval of confusion'. This interval represents the subgraph, regrouping the various concepts able to be substituted with the confusion term. On the one hand, if the terms are in a direct commutation with this last one, the interval is qualified as an 'immediate interval'. On the other hand, if the subgraph is composed of terms in an indirect commutation, the interval is 'indirect' (Boulesnane, 2008).

The immediate interval exploitation allows the consulting actors to start the information research and needs' analysis. This first level contains the contextual relations commuted immediately with the term considered. The transition to the indirect interval is only possible if the user wants to refine the research (users' questions).

Communication of Information

Furthermore, the use of the knowledge-based system herein proposed depends on the tools of information representation and exploitation, but the human expert represents the ultimate decision-maker who can orient the analysis of information.

The system is based on a heuristics method, which considers the potential ways to be followed in the situation of a problem resolution. The fundamental object is not to guarantee the most efficacious way, but to assist the user by proposing the potential case of confusion, which they can follow to resolve the problem. The consulting actors have several options and can contract these ways progressively, to refine the customers' needs (Boulesnane and Bouzidi, 2009). The choice of a relevant and adequate interpretation allows the actors using the system to validate the customers' needs (choice of interpretation).

The global system functioning is summarized in Figure 3.

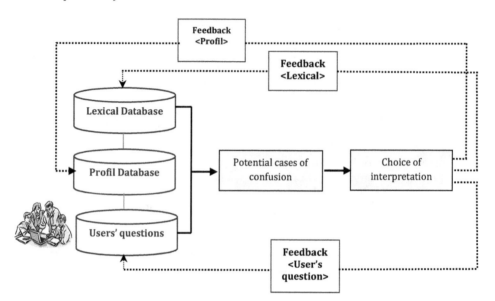

Fig 3. Global System Functioning

The case-based reasoning is pertinent in this research context (Bergmann et al, 2003). Indeed, a first phase of indexing allows the verification of whether the case which appears already exists in the system (lexical, profile and users' question feedbacks).

Among the actors involved in our approach, this system can be very useful for junior

consultants. The collaboration between the actors, who intervene in the process of information intelligence, is made in respect of business ethics which govern this professional practice.

The skills required for the implementation of the information intelligence is represented in the table below.

Table 1. Information Intelligence Process

Information intelligence's phases	Processes description
Information Collection	Analysis of documents (corpus constitution)
Information Exploitation	Semantic conceptual relation (exploitation strategies)
Information Management	Immediate and indirect intervals (cases of confusion analysis)
Information Communication	Global system functioning (feedback)

Result and Discussion

To exploit the present approach, we lean on database management systems (relational databases). The exploitation of the system

leans on the complementary phase (Figure 4).

The 'Primary information analysis' phase allows, from the documents' analysis, to establish separately the information characterizing the profiles identification and the information representing the needs' formulation

The 'primary information structuring' results in the constitution of a Lexical and Profile Database.

The phase of 'Information exploitation' allows the formalization of the process, represented by a succession of actions. In the treatment phase, the system's users can

access to every constituent of the system. They have in particular the possibility to exploit the Lexical and Profile databases (concepts, semantic relations and actor profiles). The access to the indexation of needs and profile is also possible.

Finally, the 'information needs' indexation' is represented by the formalization of the hybrid approach. This phase allows indexing information needs and profile information, in order to supply the different databases.

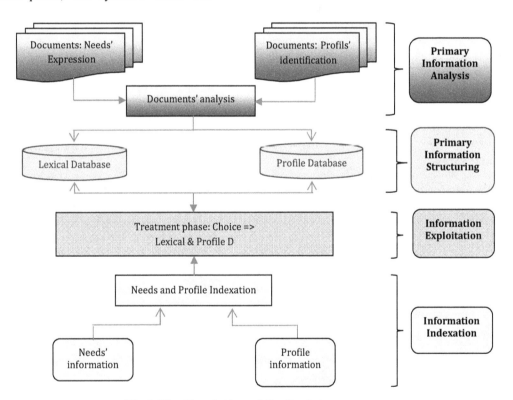

Fig 4. The Simulation of the Prototype

Research Limitation and Perspectives

The experimentation of the system was applied to an audit and advisory company, specialising in information technologies. About ten document sources, summarizing ambiguous needs, were exploited in this experiment.

Different elements allow a summary of the feedback, generated from the system experimentation. We cite, for example, the fact that the involvement of all the co-workers for this approach is important. The relevance of the used approach depends on the motivation of the users of system.

The various feedbacks allowed identifying some improvements. We list, for example, the extension of the technological lexicon, the improvement of the prototype, at both conceptual and technical levels, and the proposition of a practical guide allowing the clarification of our proposal. These various improvements constitute the research perspectives.

Conclusion

The approach proposed in this article makes a contribution to the resolution of the problems of needs' interpretation. The basis of the process of information intelligence is built around an Assistance System of needs' interpretation. This approach allows the recommendation of appropriate and pertinent technological solutions. The experiments of the system allowed identifying the research perspective.

References

Bellon, B. (2002). "Quelques Fondements de l'intelligence Economique," *Revue d'économie industrielle*, Vol 98, n° 98. pp. 55-74.

Bergmann, R., Althoff, K.- D., Breen, S., Göker, M., Manago, M. Traphoner, R. & Wess, S. (2003). "Developing Industrial Case-Based Reasoning Applications," Germany: *Springer*. 236 p. ISBN: 3-540-20733-6.

Boisvert, D. (2010). Le Développement de L'intelligence Informationnelle: Les Acteurs, les Défis et la Quête de Sens, *Editions ASTED*.

Boulesnane, S. (2008). 'Proposition d'une Approche de Médiation pour l'aide à l'interprétation des Besoins Informationnels: Contexte d'audit et de Conseil en Systèmes d'Information et en Technologies de l'Information et de la Communication,' *Thèse de Doctorat*, Université Jean Moulin-Lyon 3.

Boulesnane, S. & Bouzidi, L. (2009). 'Système d'aide à l'interprétation des Besoins: Vers une Approche Hybride,' *Revue Les Cahiers du Numérique. N° spécial en Intelligence Economique*. v.5, n°4, p.138-164. Edition : Lavoisier.

Crowne, K. A. (2009). "The Relationships among Social Intelligence, Emotional Intelligence and Cultural Intelligence," *Organization Management Journal*; Fall2009, Vol. 6 Issue 3, p148-163, 16p.

DeWalt, K. M., DeWalt, B. R. & Wayland, C. B. (1998). 'Participant Observation,' In H.R. Bernard (Ed.), Handbook of methods in cultural anthropology. Pp: 259-299. Walnut Creek, CA: *AltaMira Press*.

Ermine, J.- L. (2003). "La Gestion des Connaissances," 166p. Edition: *Lavoisier*.

Harris, M. D. S., Herron, D. & Iwanicki, S. (2008). The Business Value of IT: Managing Risks, Optimizing Performance, and Measuring Result, Édition: Boca Raton, FL: *CRC Press*, ISBN : 978-1-4200-6474-2 ; 1420064746.

Hauch, R., Miller, A. & Cardwell, R. (2005). "Information Intelligence: Metadata for Information Discovery, Access, and Integration," SIGMOD '05 Proceedings of the 2005 ACM SIGMOD international conference on Management of data. *ACM* New York, NY, USA. ISBN: 1-59593-060-4 doi>10.1145/1066157.1066250.

Laudon, K. C. & Laudon, J. P. (2011). "Management Information Systems," *MyMISLab Series*. Edition 12, illustrée. Prentice Hall PTR, 2011. ISBN 0132142856, 9780132142854. 557p.

Nanoka, I. (1994). "A Dynamic Theory of Organizational Knowledge Creation," *Organization Science*, 5(1), 14–37.

Poirier, D. (2000). 'L'intelligence Informationnelle du Chercheur: Compétences Requises à l'ère du Virtuel,' Québec: *Bibliothèque de l'Université Laval*. [Online], [January 19, 2012], Available: http://www.4.bibl.ulaval.ca/poirier/intelligence_informationnelle/definition.htm>.

Sabrina, B. & Laid, B. (2009). "Formulation des Besoins Informationnels dans une Activité Complexe et Dynamique: L'audit et le Conseil en Système d'Information et Nouvelles Technologies," of the *IBIMA (CIBIMA) journal of the International Business Information Management Association.* V.10, n°10. p. 72-84Communications [Online], [Retrieved Mai 03, 2010],

http://www.ibimapublishing.com/journals/ CIBIMA/volume10/v10n10.pdf

Šikýř, M. (2010). 'Human Resource Management Best Practices in Managing Knowledge Workers,' *In Theory management* 2., 79 – 84, Žilina.

Silverman, D. (2009). Doing Qualitative Research, London: *Sage Publications.* 456

17

Factors Driving Knowledge Creation among Private Sector Organizations: Empirical Evidence from Malaysia

Lee Wai Yi[1] and Sharmila Jayasingam[2]

[1]UNDP Malaysia, Kuala Lumpur, Malaysia

[2]Faculty of Business and Accountancy, University of Malaya, Malaysia

Abstract

The Tenth Malaysia Plan (2011 – 2015) [Tenth Plan] mentioned that in a quest to move from a middle income nation to high income nation, it is crucial for Malaysia to focus on innovation and knowledge-based growth. Despite the increasing attention focused upon knowledge management, particularly in the area of knowledge creation/innovation in Malaysia, organizations have yet to achieve the desired level of knowledge creation. Therefore, this study aims to determine what factors will influence knowledge creation among private sector organizations in Malaysia. This research examined how these four factors -- organization culture (sharing culture), organization structure (restrictive structure), Information Communication Technologies (ICT) and Human capital-- influences knowledge creation. The social system within the organization which includes knowledge sharing culture and human capital were found to positively influence the extent of knowledge creation. A restrictive organization structure had unexpected effect on knowledge creation whereas ICT was found to be only an enabler and not a driving factor.

Keywords: Knowledge creation, Malaysia, Knowledge sharing culture, Organization structure, Information Communication Technologies, Human capital.

Introduction

Since the early 1990s, Malaysia began paving the path and laying the foundation for its knowledge-based economy (k-economy) bearing the notion that continual effort is needed to improve the nation's and it's industries' competitive position (EPU, 2004; EPU, 2009). However, the initiatives only began to gain notable momentum in the late 1990s and early 2000 with the establishment of the Multimedia Super Corridor and its flagships (1996), the Third Outline Prospective Plan (2001), and the Knowledge Economy Master Plan.

Knowledge Management (KM) initiatives were also set up at various government organizations (e.g. INTAN, MAMPU, MINT, SIRIM, Telekom Malaysia, TNB), educational institutions (e.g. Multimedia University, Universiti Putra Malaysia (UPM), Open University Malaysia (OUM), Monash University (Malaysia), Universiti Teknologi Mara (UITM)), and even financial institutions (e.g. CIMB, OCBC, Bank Mualamat) (Chowdry,

2006). The measures undertaken by such institutions as mentioned above focuses on the development of knowledge enablers such as the development of human capital, research and development, information and communication technology, infrastructure and info-structure, and so on.

Evidently, concerted effort is in place to help the transformations of organizations to become knowledge-intensive firms. However, even with all these combined efforts by the Government, the impact and actual results of these initiatives are yet to be seen. In spite of the increasing attention showered upon KM, organizations have yet to achieve the desired level of KM especially in terms of knowledge creation.

Given the situation, one might think there must be some progress, especially in terms of knowledge creation. Conversely, a survey conducted by EPU (2004) on 1819 organizations from 18 industries found that despite the numerous initiatives in place, Malaysia was reported to be still lagging behind leading economies such as the United States and Singapore with regards to knowledge enablers specifically in terms of educated population, the number of computers, and the number of internet users. Malaysia was reported to be almost at par with developed nations only in terms of technological cooperation. Other attempts to evaluate KM, led researchers (e.g. Rahman, 2004; Toh, Jantan, and Ramayah, 2003; Chong, 2006) to report that the implementation of KM was still relatively slow in the Malaysian context. Although most organizations were aware of KM and its impending benefits, Chong (2006) found that the level of implementation was not at par with the level of awareness.

Furthermore, it has been reported that there is a wide discrepancy in the level of KM practices in Malaysia when compared to leading economies and foreign owned firms

(EPU, 2004). Although the second phase of the Knowledge Content Survey reported that the extent of knowledge enablers such as human capabilities, knowledge leadership, technology/info structures, and knowledge environment has improved across industries since the first survey, a decline was noted in the level of knowledge processes (knowledge generation, acquisition, sharing, and utilization) (EPU, 2009). In fact, most Malaysian firms leaned towards knowledge acquisition through hiring and shied away from actual knowledge acquisition (Jayasingam, et al., 2012).

Despite these drawbacks, the government of Malaysia is persistent in its quest to become a knowledge-intensive nation. The importance placed on k-economy by the nation is evident even in the recently introduced Tenth Malaysia Plan (2011 – 2015) (Tenth Plan). The fundamental themes of the Tenth Plan have been demarcated as the 10 Big Ideas. The essence of the 10 Big Ideas clearly delineates the need for the nation to unleash its innovative capabilities. Increased attention is being showered upon the development of soft infrastructure such as the development of human capital. A shift in focus from a capital intensive economy to a knowledge-intensive and innovation-led economy is aimed at facilitating the country's quest in achieving the status of a high income nation.

Therefore, we believe, firms need to make the transition from being good adopters and adaptors of technology to being good innovators—in other words, knowledge creators. Hence, this study aims to determine what factors will influence knowledge creation among private sector organizations in Malaysia. This research will focus on how factors such as organization culture (sharing culture), organization structure (restrictive structure), Information Communication Technologies (ICT) and Human capital will influence knowledge creation.

Theory and Hypotheses

Knowledge: The Essence of Competitive Advantage

The core ingredient for KM is knowledge. Knowledge exists at a higher order than information (Ahmed, Lim, and Zairi, 1999). Contrary to information which merely supplies facts in a structured outline, knowledge allows for making predictions, causal associations, or predictive decisions about what to do (Tiwana, 2003). Knowledge is a mix of experience, values, related information, expert insight and grounded intuition that provides an environment and framework for evaluating and incorporating new experiences and information (Awad and Ghaziri, 2004; Davenport and Prusak, 1998). In short, knowledge is data and information that has been altered into a more significant form with the influence of personal belief, value and experience (Beijerse, 1999; Beveren, 2002; Zolingen, Streumer, and Stooker, 2001).

KM has evolved as a strategic process that has a clear link to organizational performance (Jayasingam et al., 2012). Most organizations are seeking benefits of KM in order to build on their competitive advantage such as gathering and sharing best practices, effectively managing customer relationships and delivering competitive intelligence (Ming Yu, 2002; Syed-Ikhsan and Rowland, 2004). Attempts to reap the promised benefits are made through active engagement in various KM practices such as knowledge acquisition, knowledge sharing, and knowledge creation. Knowledge acquisition relates to the discovery of knowledge, (Darroch, 2003). After acquiring knowledge, one might explore the idea of knowledge sharing. This practice refers to the exchange of knowledge between at least two parties in a mutual process allowing restructuring and sense-making of the knowledge in novel milieus (Willem, 2003). A chronological order stemming from knowledge sharing would be knowledge creation or innovation. Some might refer to it as knowledge generation.

For purposes of this study, the term will be referred to as knowledge creation.

Knowledge Creation: A Strategic Tool

Over the years, KM has been acknowledged as a strategic tool to establish competitive advantage. However, a large number of researches focused on KM practices such as knowledge sharing and transfer and paid less attention to knowledge creation (Mitchell and Boyle, 2010). Challenges associated with the measurement and conceptualization of the term knowledge creation was identified as the main reason for the dearth of research related to this practice. For the purpose of this study, we adopted the definition of knowledge creation provided by Michell and Boyle (2010), which delineates knowledge creation as "the generation, development, implementation, and exploitation of new ideas" (p. 70).

In Malaysia, the trend indicates that a growing number of firms are actively promoting knowledge creation (EPU, 2005). Despite this increase in awareness and encouragement, patent and copyright applications--the two very important variables to measure innovation and knowledge creation, continue to show slow growth. Only 1815 patents and Intellectual Property (IP) rights have since been registered in the past 12 years.

On a separate note, local firms did not even make it to the Global Most Admired Knowledge Enterprises (MAKE) Winners' listing. For example, in MAKE 2006's list, North America took top honours with 8 winners, followed by Asia with 6 winners and Europe with 2 winners. Although Asia came in second place, Malaysia was not amongst the countries that won. Instead these countries were Japan, South Korea and India. Again from 2007 to 2010, Asian MAKE winners were from Indonesia, India, Japan, South Korea and Singapore. Unfortunately, Malaysia still failed to make the list.

Although the MAKE list may not be an absolute measure of the extent on knowledge creation in Malaysia, it does compel us to reflect on where Malaysian firms went wrong. The criteria used for nominating the leading knowledge driven organizations focuses on knowledge creation capabilities such as innovation capability, value creation capability and so forth. This revealed that indeed the levels of innovation and knowledge creation in the country are still at low levels although systems and structures are in place to support it. Firms will need to move from being just good adopters and adapters of technology to being good innovators instead as recommended by the study undertaken by EPU (2009).

Drivers of Knowledge Creation

Most researchers have focused on factors that influence the extent of KM generally and knowledge sharing specifically. Not many studies have attempted to explore the influence of various factors on knowledge creation specifically. Hence, this study will focus on the four factors that have been highlighted as the strategic focus areas in the Malaysian Knowledge Based Economy Master Plan and test its influence on knowledge creation. These four factors are organization culture (sharing culture), organization structure (restrictive structure), Information Communication Technologies (ICTs) and Human capital.

Knowledge Sharing Culture

Steyn and Kahn (2008) posit that almost all approaches to KM will regard organizational culture as one of the key, if not the key component of any effective KM strategy. Organizational culture can either drive or inhibit an organization's KM strategy. Numerous researchers (e.g., Toh *et al.*, 2003, Chong, 2006) have highlighted the importance of organizational culture in impacting KM practices. KM practices such as knowledge sharing and knowledge creation are interdependent processes (Janz and Prasarnphanich, 2003). Hence, when

knowledge is not shared and knowledge hoarding prevails in an organization's culture, knowledge creation will be hampered. Thus, we hypothesize that:

H1: A knowledge sharing culture will positively influence the level of knowledge creation.

Restrictive Organization Structure

In the Malaysian context, many organizations are still very mechanistic in nature, perhaps due to culture pertinent to the Asian region where hierarchy of authority and high power distance relations seems to be still acceptable in practice (Hofstede, 1980). However, Chong and Choi (2005) suggest that such organizational constraints lead to inefficiency, ineffectiveness and powerlessness within an organization. They tend to create hierarchical bureaucracy with few incentives to innovate. Eventually, this will lead to slow responsiveness to decision making processes. Therefore, according to Steyn and Kahn (2008), organizations will need to make a number of elemental changes in terms of organizational structure to become more project based and team oriented. The shift in structure should include moving individual work to team work, transforming functional work to project-based work, replacing single-skilled personnel with multi-skilled employees and eliminating co-ordination from above to adopt co-ordination among peers. All these seemed to suggest an organic kind of structure to improve structural integration in the organization and develop better overall creativity and innovation. Hence, we posit that:

H2: Restrictive structure will negatively influence knowledge creation activities.

Information Communication Technology (ICT)

ICTs are enablers for the knowledge creation process through the conversion of knowledge from inputs to outputs (Skyrme,

1998; Terajetgul and Charoenngam, 2006). The use of suitable ICTs facilitates data sorting and presentation, storage, flow through the organization and finally, supporting the thoughts processes that inform effective decision-making. Daud (2007) stated that for effective KM to exist, it will typically need the appropriate combination of organizational, social and managerial initiatives along with, in many cases, the deployment of appropriate technology like ICT. However, many top managers are reluctant to develop or invest in a KM program despite its vital importance due to the misconception regarding the costly nature of developing a KM system utilizing ICTs (Chong and Choi, 2005). The third hypothesis for this study is:

H3: ICT as enabling tools will positively influence knowledge creation activities.

Human Capital

EPU (2005) highlighted the importance of human capabilities, where the ability of workers is essential and paramount for them to participate actively in knowledge-intensive activities. As quoted in Tasmin and Woods (2008), KM practitioners and researchers alike tend to support the notion that KM requires the integration between IT systems or ICT and people who run the firm as means to attain innovation. ICT systems exist and can readily be available but ultimately, it is the human capital that is crucial in applying the technology and utilizing it. It can be considered that it is widely known that the most important competitive advantage to any firm is its workforce. Hence, employees and managers who are well equipped with skills and information to fulfil their responsibilities are essential success ingredient for any KM implementation (Chong and Choi, 2005). Thus, we provide the following hypothesis:

H4: Human capital will positively influence knowledge creation activities.

Method

Research Site, Participants, and Procedure

The self-administered surveys were distributed via personal contacts and networks and their extended networks. We identified respondents from knowledge-intensive firms who has access to the research and development (R and D) activities within the organization to ensure they could provide reliable data especially related to the extent of knowledge creation in the organization. Organizations operating within knowledge intensive industries such as electronics/electrical, chemical/fertilizer and services sector (Toh, Jantan, & Ramayah, 2003) were selected as the sample of this study.

We are conscious of the likelihood of common method variance due to the use of common raters to provide the measures of both the predictor (knowledge sharing culture, restrictive organization structure, ICT, and human capital) and criterion (knowledge creation) variables. The study's internal validity could probably be affected by this (Podsakoff *et al.*, 2003). Hence, to reduce the effect of common method variance, we created a psychological separation between the predictor and criterion variable as suggested by Podsakoff *et al* (2003). This step was taken to avoid the assumption among respondents that the measurement of the criterion variables is related to the predictor variable. In addition, we assured the respondents that there was no right or wrong answers and what mattered most was the respondents' honest opinion. This was done to decrease the likelihood of them trying to link the answers for the predictor and criterion variable and provide answers as probably anticipated by the researcher (Podsakoff *et al.*, 2003).

The questionnaires were either hand delivered or emailed to the potential respondents. The snowballing technique was employed when distributing the surveys where we targeted one contact on our network and in return, that one contact was requested to contact 5 others in their extended network to participate in the survey. Utilizing this method somewhat improved the speed for data collection and improved response rate.

Out of 250 questionnaires sent out through various channels, 210 were received but only 205 were usable questionnaires, giving an effective response rate of 82%. This study had a fairly proportionate distribution of male (52.7%) and female (47.3%) respondents. As for age groups, 57.6% were 35 years or younger whilst the remaining 42.4% were aged 36 and above, indicating a majority of younger respondents. Most of the respondents possessed at least a bachelor's degree and beyond at postgraduate levels of education. This indicates that the respondents were highly educated and could be due to the fact the majority of the respondents were at officers/executives, supervisory, management and senior management levels, with a cumulative total of 94.1%.

Measures

The measures were adapted from two sources – the measures for the four factors were adopted from the work of Syed-Ikhsan and Rowland (2004). The measures for knowledge creation were adapted from the Malaysian Knowledge Content Survey (EPU, 2009). Since both sources already tested the questionnaires in Malaysia, it made the pilot test unnecessary. This section utilized the 7-point Likert Scale which required the respondents to indicate their levels of agreement and disagreement by placing a "X" at the following appropriate number (1=Strongly Disagree, 2=Disagree, 3=Slightly Disagree, 4=Neutral, 5=Slightly Agree, 6=Agree and 7=Strongly Agree.)

The survey contains three main sections; the first section asked questions related to the respondent's organization's KM strategy. The second section contains thirty questions related to the respondents and the organization where they work which will measure the dimensions of sharing culture, restrictive organization structure, ICTs, human capital and finally, knowledge creation. Sample items include: "...The management provides time and resources to take part in the learning and sharing exercises " (human capital)"...All staff are ready and willing to give advice and help upon request" (organization culture); "...Computer-based information systems provide me with more up-to-date information than that available in manual files" (ICT); "...The confidentiality status of documents in my organization often leads to problems in acquiring information and creating knowledge" (organization structure); and "my organization has constantly filed new applications for patents, designs, know-how, etc. in the past one year"(knowledge creation).

Results

Psychometric Properties of Measures

Exploratory factor analysis (EFA) was conducted on the four factors of sharing culture, restrictive structure, ICT and human capital. This was done to examine the correlations between the different variables in the study and to determine whether the data could be condensed or summarized into smaller set of factors.

For the independent variables, the factors were rotated using Oblimin with Kaiser Normalization method because the factors are assumed to be related. There were 5 factors created initially but it was left with 4 factors at the end after selectively dropping the items with high cross loadings. Some of these factors were not deemed fit nor made any sense to their related factors, and thus had to be dropped. The Kaiser-Meyer-Olkin (KMO) measure of sampling adequacy was

0.836 demonstrating adequate inter-correlations, and the Bartlett's Test of Sphericity was significant (χ^2 = 2069.797, p < 0.01). The four factors were reliable with reliability coefficients above .70.

Factor analysis was also performed on knowledge creation items to ensure that all the 5 items fall into one factor only. The research results did show that all the 5 items fell into one factor and the name of the factor remained as knowledge creation. The KMO

measure of sampling adequacy was 0.820 whilst the Bartlett's Test of Sphericity was significant (χ^2 = 513.559, p < 0.01 at 0.000). The scale demonstrated high reliability with a Cronbach alpha of .87.

Descriptive statistics, correlation between the factors, and reliability coefficients for the subscales are shown in Table 1. Yin, R. K. (1989). 'Case Study Research: Design and Methods,' *Sage Publications Inc.*, USA.

Table 1: Descriptive Statistics, Correlation Coefficient, and Reliability Coefficient for Independent and Dependent Variables

	Mean	Std Dev.	Human Capital	Restrictive Structure	ICT	Sharing Culture
Human Capital	4.36	1.16	**.85**			
Restrictive Structure	3.94	1.09	-.15*	**.79**		
ICT	5.18	1.05	.41**	.12	**.84**	
Sharing Culture	4.42	1.02	.53**	-.28**	.30**	**.72**

Note: * p<0.05, ** p < 0.01; Diagonal entry shows reliability coefficients

Tests of Hypotheses

It was predicted that knowledge creation is still at low levels among most Malaysian organizations. The finding of this research confirms this fact. Respondents recorded that only 24.8% of their organizations had constantly filed new applications for patents, designs, and know-how in the past one year and only 21.5 percent recorded that the applications were successful. Although approximately 40% of the respondents found their organization to have increased the introduction of new products and improved processes in the past 1 year, this proportion was not sufficient to create new ideas that deserve recognition. Unfortunately, only

32.7% reported that their organizations had been actively involved in R and D activities in the past one year.

Multiple regression analysis was performed to determine the prediction power between the dependent variable (knowledge creation) and the multiple independent variables (sharing culture, restrictive structure, ICT and human capital). The results are as shown in Table 2. The model is found to be safe from multi-collinearity problems as the condition index values were all less than the cut-off point of 30 and the Variance Inflation Factor (VIF) which measures tolerance is less than 10 for all factors (Hair *et al.*, 2006).

We found only human capital (B = 0.433, p < 0.01) and sharing culture (B = 0.285, p < 0.01) to positively influence knowledge creation—hence, supporting *H1* and *H4*. Although *H2* states that restrictive structure will negatively influence knowledge creation, the results show otherwise. Furthermore, the effects of ICT were not significant and therefore, *H3* was also not supported.

Table 2: Results of Regression Analysis

	Dependent variable Knowledge creation
Independent variables	
Knowledge sharing culture	.28**
Restrictive organization structure	.16*
ICT	-.05
Human capital	.43**
F value	24.25
R^2	.32
Adjusted R^2	.31

* $p<0.05$, ** $p < 0.01$

Discussion

This study aimed at investigating the factors that influences the extent of knowledge creation in organizations. Given this objective, we tested four major hypothesized relationships: (a) the relationship between knowledge sharing culture and knowledge creation, (b) the relationship between restrictive organization structure and knowledge creation, (c) the relationship between ICT and knowledge creation, and finally (d) the relationship between human capital and knowledge creation. Our major findings are summarized below.

Major Findings

Several conclusions can be drawn from the results of this study. First, as hypothesized knowledge sharing culture creates an environment that facilitates knowledge creation. In line with the study by Chong (2006), culture is one of the most important factors for the success of a company especially in relation to knowledge creation and application. Basically, when sharing knowledge becomes a way of life in the organization and the culture strongly emphasizes that knowledge sharing is power, knowledge hoarding can be reduced (Chong and Choi, 2005). Hence, when employees share knowledge voluntarily, the knowledge that is being circulated is able to stimulate new ideas and thoughts—leading towards knowledge creation.

Next, although we hypothesized a negative effect of restrictive structure on the extent of knowledge creation, the results proved otherwise. Surprisingly, restrictive structure appears to be still acceptable in Malaysia. Ansari *et al* (2004) highlight two key components of culture in Malaysia—one of it being preference for hierarchy. Given a high power distance, Malaysian society is described as a platform where bureaucratic structures are still widely acceptable (Hofstede, 1994). Since knowledge creation is still in its' infant stage among most business organizations, a structure that is defined by clear rules, procedures, and policies are probably still needed to monitor and coordinate knowledge creation activities. Third, ICT has long been associated with successful KM systems. In a competitive business environment, organizations are investing huge amounts in information technology to establish a state of the art KM system and enhance their competitive advantage (Kakabadse, et al., 2003). However, despite the implementation of first-rate information technology, surveys point out that KM systems are failing at an equivalent pace as the rate of implementation (Ambrosio, 2000; Smith, et al., 2003). Organizations are fundamentally so obsessed with the notion that the success of the KM systems solely relies on technology—hence failing to acknowledge the fact that employees' acceptance and commitment towards the KM system is equally important (Coulson-Thomas, 1997; Davis, Subramaniam, and Westerberg, 2005). This could be indicative that ICT--which is available for use in organization--can only contribute towards knowledge creation when people utilize it as much as they should. The mere availability of technology is not sufficient to drive organizations to create knowledge.

Fundamentally, the importance of human capital is clearly supported in this study. Organizations must not overlook the fact that knowledge workers are the essence of the social system of KM projects (Alvesson, 2004; Ribiere and Sitar, 2003). Human capital competency plays an important role in helping them carry out their work in any situation (Teerajetgul and Charoenngam, 2006). Individual knowledge of knowledge workers lays the foundation for organizational knowledge. Hence, timely and appropriate employee training constitutes one of the key success factors for KM implementation (Chong and Choi, 2005).

Theoretical Contributions

Our study has some obvious theoretical implication. First, the social system within any organization seems to bear more significance in improving the extent of knowledge creation. An organization that possesses highly qualified employees and is characterized by a knowledge sharing culture has an added advantage when it comes to knowledge creation. ICT is important as the foundation of a KM system, but the integration of the social system with ICT is essential for the success of any knowledge creation initiative. Second, the influence of a

restrictive structure should be interpreted with caution. The unexpected positive influence this variable has on knowledge creation can be attributed to the Malaysian culture. However, it is also possible that the current extent of KM in Malaysia may require a more controlled organization structure to ensure knowledge creation activities are more synchronized with the organization's objectives

Practical Implications

This study has shed some light on the importance of the social system within an organization on knowledge creation. Hence, organizations should focus on establishing enabling environment for their people to share knowledge. The human capital should also be given more opportunities to develop their human capital to acquire more knowledge so as to be able to share it with their colleagues. When there is greater knowledge shared and attained, organizations can grow with a competitive edge against other rivals in the industry.

Limitations and Directions for Future Research

Our study has some potential limitations. First, we only considered four factors: knowledge sharing culture, restrictive structure, ICT, and human capital. Future research should attempt to identify other factors that can improve the extent of knowledge creation such as human resource practices, leadership style, and so on. Second, as our data were limited to the Malaysian context, it would be recommended that future researchers compare data from other different cultures. A comparative study would help shed some light on cultural differences, especially when interpreting the influence of a restrictive organization structure. Third, this was a cross sectional study which limited our ability to observe improvements in the extent of knowledge creation. We relied on the respondents' evaluation of the extent of knowledge creation in their organization in the past one year. Future researchers could attempt to develop a more objective measure of knowledge creation to test this model.

References

Ahmed, A. K., Lim, K. K. & Zairi, M. (1999). "Measurement Practice for Knowledge Management," *Journal of Workplace Learning*: Employee Counseling Today 8(11), 304-311.

Alvesson, M. (2004). Knowledge Work and Knowledge Intensive Firms. *Oxford University Press*, Great Britain.

Ambrosio, J. (2000). Knowledge Management Mistakes [Online]. [Retrieved November 23, 2005] Available: http:/ www. computerworld.com/industrytopics/energy/ story/ 0, 10801,46693,00.html

Ansari, M. A., Ahmad, Z. A. & Aafaqi, R. (2004). "Organizational Leadership in the Malaysian Context," In D. Tjosvold and K. Lueng (Eds.) *Leading in high growth Asia*: *Managing relationship for teamwork and change* (pp.109-138). Singapore: World Scientific Publishing Co.

Awad, E. M. & Ghaziri, H. M. (2004). 'Knowledge Management,' Upper Saddle River, NJ: *Prentice Hall*.

Chong, S. C. (2006). "KM Critical Success Factors: A Comparison of Perceived Importance vs Implementation in Malaysian ICT Companies,'"*The Learning Organization*, 13(3), 230-256.

Chong, S. C. & Choi, Y .S. (2005). 'Critical Factors for Knowledge Management Implementation Success,' *Journal of Knowledge Management Practice* [Online], [Retrieved July 23, 2006], Available: http://www.tlainc.com

Chowdry, N. (2006). Building KM in Malaysia, [Retrieved August 11, 2008] Available: http://www.kmtalk.net/article.php?story =20060727043623849

Coulson-Thomas, C. J. (1997). "The Future of the Organization: Selected Knowledge Management Issues," *Journal of Knowledge Management*. 1, 15-26.

Darroch, J. (2003), "Developing a Measure of Knowledge Management Behaviors and Practices," *Journal of Knowledge Management*, 7, 41-54.

Daud, R. A. (2007). "Knowledge Management Systems (KMS) in Organisation: A Collaborative Model for Decision Makers," [Online] [Retrieved June 28, 2011] Available: http://www.kmtalk.net/article.php?story=2 007071609353970&query=daud   ,

Davenport, T. H. & Prusak, L. (1998). Working Knowledge: How Organizations Manage What they Know. Boston: *Harvard Business School Press*.

Davis, J. G., Subramaniam, E. & Westerberg, A. W. (2005). "The Global and the Local in Knowledge Management," *Journal of Knowledge Management*, 19(1), 101-112.

EPU (2004). 'Knowledge Content in Key Economic Sectors in Malaysia,' Economic Planning Unit, Prime Minister's Department Malaysia.

EPU (2007). 'Malaysia and the Knowledge Economy: Building a World-Class Higher Education System,' [Online] [Retrieved August 8, 2008]. Available: http://siteresources.worldbank.org/INTMAL AYSIA/Resources/ Malaysia-Knowledge-Economy2007.pdf

EPU (2009). Knowledge Content in Key Economic Sectors in Malaysia-Phase II. *Economic Planning Unit, Prime Minister's Department Malaysia.*

Hofstede, G. (1980). 'Culture's Consequence: International Differences in Work Related Values,' Beverly Hills, CA: Sage.

Hofstede, G. (1994). 'Management in a Multicultural Society,' *Malaysian Management Review.* 29, 3-12.

Janz, B. D. & Prasarnphanich, P. (2003). "Understanding the Antecedents of Effective Knowledge Management: The Importance of Knowledge-Centered Culture," *Decision Sciences*, 34(2), 351-384.

Jayasingam, S., Ansari, M. A., Ramayah, T. & Jantan, M. (2012). "Knowledge Management Practices and Performance: Are They Truly Linked?," *Knowledge Management Research & Practice,* advance online publication, March 12, 2012.

Kakabadse, N. K., Kakabadse, A. & Kouzmin, A. (2003). "Reviewing the Knowledge Management Literature: Towards Taxonomy," *Journal of Knowledge Management* 7, 75-91.

Ming Yu, C. (2002). 'Socializing Knowledge Management: The Influence of the Opinion Leader,' *Journal of Knowledge Management Practice.* 3, 76-83.

Mitchell, R. & Boyle, B. (2010). "Knowledge Creation Measurement Methods," *Journal of Knowledge Management*, 14(1), 67-82.

Podsakoff, P. M., MacKenzie, S. B., Lee, J.- Y. & Podsakoff, N. P. (2003). "Common Method Biases in Behavioral Research: A Critical Review of the Literature and Recommended Remedies," *Journal of Applied Psychology.* 88, 879-903.

Rahman, B. A. (2004). 'Knowledge Management Initiatives: Exploratory Study in Malaysia,' *Journal of American Academy of Business.* 4, 330-336.

Ribiere, V. M. & Sitar, A. S. (2003). "Critical Role of Leadership in Nurturing a Knowledge-Supporting Culture," *Knowledge Management Research & Practice.* 1, 39-48.

Skyrme, D. (1997). "Knowledge Management: Making Sense of an Oxymoron," [Online], [Retrieved July 5, 2005], Available: http://www.skyrme.com/insights/22km.htm

Smith, G., Blackman, D. & Good, B. (2003). "Knowledge Sharing and Organizational Learning: The Impact of Social Architecture at Ordnance Survey," *Journal of Information and Knowledge Management Practice*, [Online] 4, [Retrieved20 December, 2004] Available: http://www.tlainc.com/articl50.htm

Steyn, C. & Kahn, M. (2008). "Toward the Development of a Knowledge Management Practices Survey for Application in Knowledge Intensive Organisations," *South African Journal of Business Management.* 39(1), 45-53

Syed-Ikhsan, S. O. S. & Rowland, F. (2004). "Benchmarking Knowledge Management in a Public Organization in Malaysia," *Benchmarking: An International Journal*, 11(3), 238-266.

Tasmin, R. & Woods, P. (2008). "Knowledge Management and Innovation in Peninsular Malaysia," Proceedings of KMICe 2008 conference. 10-12 June 2008, Langkawi, Malaysia.

Teerajetgul, W. & Charoenngam, C. (2006). "Factors Inducing Knowledge Creation: Empirical Evidence from Thai Construction Projects," *Engineering, Construction and Architectural Management*, 3(16), 584-599.

Tiwana, A. (2003). 'The knowledge Management Toolkit: Orchestrating IT, Strategy and Knowledge Platforms,' *Prentice Hall*, Upper Saddle River, NJ.

Toh, H. H., Jantan, M. & Ramayah, T. (2003). 'Knowledge Management: An Exploratory Study on Malaysian Organization,' *The*

International Journal of Knowledge, Culture and Change Management, 3, 995-1014.

Uit Beijerse, R. P. (1999). "Questions in Knowledge Management: Defining and Conceptualizing a Phenomenon," *Journal of Knowledge Management.* 3(2), 94-109.

Van Beveren, J. (2002). "A Model of Knowledge Acquisition that Refocuses Knowledge Management," *Journal of Knowledge Management.* 6(1), 18-22.

van Zolingen, S. J., Streumer, J. N. & Stooker, M. (2001). "Problems in knowledge Management: A Case Study of Knowledge-Intensive Company," *International Journal of Training and Development*, 5, 168-184.

Willem, A. (2003). "The Role of Organisation Specific Integration Mechanisms in Inter-Unit Knowledge Sharing," Vlerick Leuven Gent Management School, Ghent University, Belgium [Online PhD Dissertation], [Retrieved June 17, 2004] Available: http:72.14.203.104/search?q=cache:AwAf_o k1x7UJ:www.ofenhandwerk.com/oklc/pdf

Permissions

The contributors of this book come from diverse backgrounds, making this book a truly international effort. This book will bring forth new frontiers with its revolutionizing research information and detailed analysis of the nascent developments around the world.

We would like to thank all the contributing authors for lending their expertise to make the book truly unique. They have played a crucial role in the development of this book. Without their invaluable contributions this book wouldn't have been possible. They have made vital efforts to compile up to date information on the varied aspects of this subject to make this book a valuable addition to the collection of many professionals and students.

This book was conceptualized with the vision of imparting up-to-date information and advanced data in this field. To ensure the same, a matchless editorial board was set up. Every individual on the board went through rigorous rounds of assessment to prove their worth. After which they invested a large part of their time researching and compiling the most relevant data for our readers.

The editorial board has been involved in producing this book since its inception. They have spent rigorous hours researching and exploring the diverse topics which have resulted in the successful publishing of this book. They have passed on their knowledge of decades through this book. To expedite this challenging task, the publisher supported the team at every step. A small team of assistant editors was also appointed to further simplify the editing procedure and attain best results for the readers.

Apart from the editorial board, the designing team has also invested a significant amount of their time in understanding the subject and creating the most relevant covers. They scrutinized every image to scout for the most suitable representation of the subject and create an appropriate cover for the book.

The publishing team has been an ardent support to the editorial, designing and production team. Their endless efforts to recruit the best for this project, has resulted in the accomplishment of this book. They are a veteran in the field of academics and their pool of knowledge is as vast as their experience in printing. Their expertise and guidance has proved useful at every step. Their uncompromising quality standards have made this book an exceptional effort. Their encouragement from time to time has been an inspiration for everyone.

The publisher and the editorial board hope that this book will prove to be a valuable piece of knowledge for researchers, students, practitioners and scholars across the globe.

List of Contributors

Mohamad Noorman Masrek
Accounting Research Institute/Faculty of Information Management, Universiti Teknologi MARA, Shah Alam, Malaysia

Nur Izzati Yusof, Siti Arpah Noordin and Rusnah Johare
Faculty of Information Management, Universiti Teknologi MARA, Shah Alam, Malaysia

Reza Sigari Tabrizi and Yeap Peik Foong
Multimedia University, Cyberjaya, Malaysia

Nazli Ebrahimi
University of Malaya, KL, Malaysia

Abrar Haider
School of Computer and Information Science, University of South Australia, Australia

Abdelfatteh Triki and Fekhta Zouaoui
Institut Supérieur de Gestion- Tunis

Akmal Aris, Juhana Salim, Shahrul Azman Mohd Noah and Kamsuriah Ahmad
Universiti Kebangsaan Malaysia, Bangi, Malaysia

Rosângela Formentini Caldas
São Paulo State University, Information Science Department, São Paulo, Brazil

Mohamad Noorman Masrek, Hasnah Abdul Rahim, Rusnah Johare and Yanti Rahayu Rambli
Universiti Teknologi MARA, Shah Alam, Malaysia

Peter Kažimír and Vladimír Bureš
College of Management, Bratislava, Slovakia

Vladimír Bureš and Tereza Otčenášková
University of Hradec Králové, Hradec Králové, Czech Republic

Barbora Jetmarová
University of Pardubice, Faculty of Economics and Administration, Pardubice, Czech Republic

Tereza Otčenášková, Vladimír Bureš and Jaroslava Mikulecká
University of Hradec Králové, Hradec Králové, Czech Republic

Umar Ruhi
Telfer School of Management, University of Ottawa, Ottawa, Canada

Dina Al-Mohsen
E-Business Technologies, University of Ottawa, Ottawa, Canada

Afaf Mubarak
Faculty of Business Administration1, Al Hosn university- Abu Dhabi- United Arab Emirates

Kamla Ali Al-Busaidi
Sultan Qaboos University, Alkhod, Oman

Anushia Chelvarayan, Chandrika Mohd Jayothisa and Hazlaili Hashim
Centre For Diploma Programme, Multimedia Universiti, Jalan Ayer Keroh Lama, Melaka

Khairol Nizat Lajis
Foundation Studies and Extension Education, Multimedia Universiti, Jalan Ayer Keroh Lama, Melaka

Amir Raslan Abu Bakar
Islamic Development Bank, Jeddah, Saudi Arabia

Rugayah Hashim
Universiti Teknlologi Mara, Shah Alam, Selangor, Malaysia

Boulesnane Sabrina, Bouzidi Laïd and Marini Jean-Luc
Jean Moulin - Lyon 3 University, Lyon. France

Lee Wai Yi
UNDP Malaysia, Kuala Lumpur, Malaysia

Sharmila Jayasingam
Faculty of Business and Accountancy, University of Malaya, Malaysia

Printed in the USA
CPSIA information can be obtained
at www.ICGtesting.com
JSHW051443221024
72173JS00006B/1565